Excuse Me
Life is Calling

CHRISTIAN PEDERSEN

FriesenPress

Suite 300 - 990 Fort St
Victoria, BC, V8V 3K2
Canada

www.friesenpress.com

ISBN
978-1-5255-4231-2 (Hardcover)
978-1-5255-4232-9 (Paperback)
978-1-5255-4233-6 (eBook)

1. Religion, Religion & Science

Distributed to the trade by The Ingram Book Company

To learn, one must not only read, but must also think.
Thinking is best done through discussion.

TABLE OF CONTENTS

PRELUDE

Suddenly, the world is facing the loss of liberalism and perhaps even democracy.

The United Kingdom has voted to leave the European Union, likely because of the hordes of immigrants desirous of gaining access to their very beneficial social programs—programs not seen in other countries. Canada is potentially the second most desirous for immigrants and refugees because of its generous benefits. The Scandinavian countries and Holland used to be the most favorable countries for immigration, but they, too, have changed some of their generosity toward Third-World economic refugees.

The voters of the United States have decided that they want a president who favors protectionist policies; one who limits immigration only to those who meet certain criteria desired by the mainstream electoral population; one who also openly demonstrates disdain for women and calls foreign nations "shit hole" countries.

The world's two historically outstanding proponents of democracy, immigration, liberty, and equality of man—the UK and the USA—have closed shop. There are no other viable options for the world's population; options offering equality and an opportunity to live a good life without duress. Sure, there are other forms of government—such as those found in Russia, Saudi Arabia, Iran, Iraq, Afghanistan, the Islamic Republic of Pakistan, and so on—but you don't see hordes of refugees flocking to gain access to them.

Democracy is dying. No political system in history has demonstrated that it can withstand the test of time. Communist Russia, dictatorships, theocracies, feudalist states, kingdoms, and so on, all failed long before this. Democracy is the last to fall.

Opportunities for big business and/or artificial intelligence to take over are paramount. Computer modeling is the new intelligence. Data about everyone on this planet has been and is being collected in vast amounts. This field of new science can affect the way you shop, portending a new way to influence you in your daily life. Japan already uses artificial intelligence robots in the home care industry to great fanfare and with positive results.

How do we, the peasants of the world, clearly understand what might happen and where this might lead? Can we discern the hidden agenda of those who would control us? Do they proffer truthful insight into our daily living?

History repeats itself; that is also precisely the precept that computer modeling understands. We need to recognize the two greatest building blocks of civilization in history to access our options critically and understand what will be happening in the very near future. Theology and science have created an understanding within our species and built a set of cultures that have enabled us to build civilization without destroying each other, with obvious exceptions.

This book is about the critical assessment of theology and science, how the two have intermingled and been misunderstood across the ages, how science today has taken on attributes of religious fervor, and how religion stated for hundreds of years that it was the interpreter of science. In the end, science is just science, and theology is just religion. Both are very important in their own right, yet each is different from the other. Both have built cultures that nurture and support vast populations. Both have worked together to build a modern society that has worked for a very long time. Until now.

We need to live in a world where we ask questions that give answers, not in a world where the answers cannot be questioned. To that end, I have included a series of questions at the end of every chapter. I invite you to discuss these questions with someone else present—especially a theologian or scientist, or both, if possible. It is you who needs to make up your mind about the truth of our lives and what is available to us in the future.

To read is to understand; to discuss is to learn. To learn, one must not only read, but must also think. Thinking is best done through discussion.

Christian Pedersen

INTRODUCTION

As science gives us more and more insight into the creation of our world and the universe, I ask myself: Is there only science, or only God in all this, and which or both, and where do they intersect? If they even intersect? Why do so many scientists deny the existence of God, while many others are openly skeptical? Scientific thinking has an audience that, like sports fans, jumps on the proverbial bandwagon with support for the winning team. I respect scientific thinking and value the genius and knowledge gained by scientific study that is so freely given to us. I also feel scientists have been chosen to do this work and have been allowed to complete it in this time as an act of a loving God who is walking with us through this process and allowing us to learn his works as we can accept them as humans on this planet. That God wants us to understand his role and accept his hand in the development of the universe as he discloses it to us.

Mankind, from the spread of *Homo sapiens* throughout the world 80,000 years ago to present day, has generally spoken of a deity. Yes, several philosophers in Greece and other places were contrarian to this thought, but theism mostly won out.

I have explored this in my own way, reading extensively on topics that were of interest and were, hopefully, enlightening. I needed to know what theism is and all its versions. Is theism the same as theology, since numerous theologians have urged me not to use the terms interchangeably?

Where did all the God stuff come from, and how was it the same yet also different in all cultures? A person could spend several lifetimes on this subject alone and still not be satisfied that they were cognizant of all there is to know. Simplifying this does not do justice to this subject. But that is precisely what I have done here—not to understate the complexity of this subject but rather to make it somewhat understandable to those of us who want a cursory understanding of the subject.

I strove to just look at the numerous main religions from which several variants have evolved. We could discuss the variants forever, and it mostly becomes opinion and whose opinion is correct. There is so much excellent material on this subject that I have just glossed over much of it and added my own perspective.

I am also trying to present the version of theology that is presented by the founders of each religion. Having extensively read a lot of literature on all subjects, I have included words, phrases, and even sentences that I gleaned from remembering the words of each philosopher of each religion's practice. This is my effort to present each religious practice as fairly and closely to the intention of the leadership of each religion as possible. It is not intended to interpret according to my own layperson understanding.

My own background is Evangelical Lutheran, so I often lean on my own understanding from this perspective. Also, within Lutheranism, the theologians, many of whom are friends whom I deeply respect, are highly trained experts within its specific theology. The fantastic thing about being an Evangelical Lutheran is that I can hold opinions of religion that may be abhorrent to others and still be accepted by the Lutheran leadership as a thinking, believing person, if I accept the basic tenet of believing in Christ.

At the same time, I have tried to do my best to represent what I thought I was hearing from pastors and my learned friends. I asked them to review my thoughts with respect to their individual areas of expertise. These friends accommodated me with patience, understanding, and grace. Some even introduced me to their respected teachers, leaders, and professors with the idea of gleaning better insight into my subjects. I am not a scientist; I am just an observer with the intention of learning and providing my insight to others. The elite of the educated theologians I met gave lip service to my questions, and within one hundred words, I determined that I was being given all the accommodation that a housewife would give to dust under the rug. This attitude, of course, is an occupational hazard of many learned professionals educated to the pinnacle of their profession; thus, the brush-off to avoid further interaction with those who are less trained. Sadly, the dichotomy of all this is the statement that "knowledge will change everything," but at the same time, our inherent personalities are only positioned to criticize those who are not our equals.

We need to understand how we became the alpha of all species, how we relate to each other, and why we can and do interrelate to each in communion with our entire species. I go into this subject with opinions obtained from great teachers and historians, complemented with private thoughts and a determination to engage you in talking about the subject. Often, I will place on the pages something that is controversial to cause you to stop, think, and discuss what is being said. It would be significantly constructive if you engaged the services of an educated theologian of your faith when you read dissertations of theology as I have represented them in this manuscript. All the subjects in this book are meant to be explored with someone who is trained and can impart further clarification or change the direction of your thought. This is not a book about scientific discovery so much as a book of revelation for those of you who have an interest in today's world and our future on this planet.

For me to attempt to interpret the teachings of those who have spent a lifetime studying, it became imperative that I understood by reading the works of the learned about the role of man in the historical development of Earth with *Homo sapiens* becoming the leaders of the animal kingdom. How did we populate this planet? How were we created? What is our timeline of development, and what is the development that made us so superior, in our own minds?

With that process in mind, it becomes apparent that our evolution on Earth is part of our DNA. Since our genome has recently been solved, along with that of most other life forms, I needed to understand how this impacts humans. So, there is a volume on medicine and DNA, its history, how it affects and has affected us, and how it will change the future of this planet. How this will be used, either for better or for worse, is debatable.

The stars, as civilization has seen and interpreted them over the centuries, are today washed out of our minds because of the ambient light within our cities, but throughout history, they were the guideposts to living. The stars represent a discernment of ancient principles that were dictated regarding the construction of the universe as they saw it, often including divine intervention. Everything was determined to be true by the "truth" of the eye. Theology was deemed to be part of the essence of the sky and overlapped science, creating roadblocks to understanding. We look at how this finally unfolded into two separate processes: science and theology. One is a factual scientific reconciliation of the processes affecting *Homo sapiens*, and the other is a process of building a great and creative environment of thought that allows man to live in harmony with faith, love, and charity.

Theology tries to tell us why something happens or what our role is within it. Science tries to tell us how something happens, through proof and logic. Philosophy is the discussion around both fields of endeavor, trying to draw conclusions without the necessity of injecting facts or proof. All three coexist in conjunction with each other. All through the ages, the three have been necessary in the process of discovery, and each is and has been regularly confused with the others as to whether science is theology or theology is science. Logical thinking requires insight into the process of discerning whether it is scientific or theological thinking, and that is not often easily discerned. The discernment requires a multi-level assessment, which we average *Homo sapiens* often seem to discard and become focused upon single facets of discovery, leaving out levels of determination. In other words, we want everything to be simple black or white.

Then, finally, the science. I will try to engage you to question science before accepting the "scientific" truth. First, the science of the stars. Along with this is the science of quanta, physics, Newton, Einstein, and the big bang as it is understood in the quanta; the big bang, theories of loop quantum gravity, dark matter, dark energy, and spacetime. This is all very foreboding and frightening to the uneducated reader. I will try to deal with all this in a way that everyone can understand, in an unpretentious and uncomplicated way. My brother is an astrophysicist by training, and when I ask him for relevance of meaning of a subject, he will often respond, "That is a little bit of Harry Potter!" What he means is that although it is very interesting, it is just something from the fantastical mind of a very creative person. Physicists also want to be understood by those of us with whom they must occasionally dialog, even if it is only to order a cup of coffee.

All this discussion leads to the subject of philosophy. How do we take an opinion and turn it into an unbiased fact? Opinion also leads to language use, or rather the language of the technology of the chosen subject—especially by PhD thinkers who are trying to get their points across to us. They invent their own vernacular. By comparison, think of two airline pilots discussing an approach to a difficult airport, with bad weather involved. They use terminology of fact with words of meaning that elude all but the trained person in that occupation. The very same for God talk, church talk, and so on. Already, in the above descriptions, we have used several words like "evangelical" and "theism". Only those trained in religious thought generally understand the meaning of these words.

"Take philosophy—please!" as a comedian would say. My reading on this subject has been the most difficult subject to comprehend, given the difficulty in understanding the basic progression of thought, of anything I have encountered in my lifetime. You get a PhD in most subjects, and as far as I can see, you have earned it by the time you get to the defense. I understand that every student working on a PhD must study René Descartes, among many others. He influenced Sir Isaac Newton, for instance.

I think therefore I am!

By the word "God" I understand a substance that is infinite, eternal, immutable, independent, supremely intelligent, supremely powerful, and which created both myself and everything else (if anything else there be) that exists. All these attributes are such that, the more carefully I concentrate on them, the less possible it seems that they could have originated from me alone. So, from what has been said, it must be concluded that God necessarily exists.[1]

Rene Descartes, philosopher,
mathematician, and scientist (1596–1650)

Quantum physics, astrophysics, religious studies, medical science, politics, and culture are all studies in philosophy at the PhD level, and the PhD designation reflects that—"Doctor of Philosophy". In one sentence, it can be said that "I love it and hate it all at the same time," while I read the incantations of those who give us their thoughts.

Scientists speak their own language. They try to make it simple, in their own way, for the rest of us to understand, but mostly it is still so elevated that we must work hard to understand. I read one book on quantum physics about seven times and still find new ideas to draw me into reading it again. So, I will try to explain this new

---

[1] Rene Descartes, *Meditations on First Philosophy with Selections from the Objections and Replies*, ed. John Cottingham (Cambridge: Cambridge UP, 1996), 31.

language for those of us who are unfamiliar with their terminology. Yes, mathematics is the universal language, as it communicates equally to all those who understand it in a clear and precise way. But this is not about mathematics; it is about the philosophy of the subject and how it might relate to God.

And then I must immediately delve into philosophical definition. Yuck! It is imperative that we understand how to discuss our ideas in a true and basic way, without all the conditioned ideas we each have from our personal historical perspectives. Philosophy uses the terms "epistemology" and "ontology". Epistemology is knowledge as perceived by people, whereas ontology is actual knowledge without the opinion of a person influencing it.

Think about this for a minute. Both are very important ways of visualizing how we deal with ideas. I have heard it said that the discussion of science and theology is thinking with facts, while thinking without facts to impede the outcome is philosophy. As I try to conclude with this dissertation, all three are vital to our species and are often intermixed, but they are necessarily independent of each other while depending upon each other for sustenance. Just like the parts of an atom.

My goal is to bring you the science and try to represent the possibility of theological belief that would be a way of understanding, backing, and supporting that science. Second, I intend to show you science that is closer to theology than science, especially if you must have faith to believe it. Third, I wish to identify some of the areas where we *Homo sapiens* have taken over the science or theology for gain or for the lust of power and control.

But first, we must understand theology, its history, various understandings, and progression through the ages. I am also going to ask you where you think it is that the hand of man, not the divine direction of God, has evolved our theism, and what is the result of that today? There will be a series of questions and places within the chapters where you will be asked to stop and ask yourself these questions, and which will perhaps promote group discussion. It is up to you to make up your own mind, as I do not propose to provide any "correct" answers. I do propose, however, to provide information that is often worthy of controversy and discussion, regardless of my own personal beliefs. I also hope to give you tools to think about how to discern what is correct in statements our new science provides, with particular regard for the new "environmental" science, which has all the possibilities of becoming the new religion.

I hope you enjoy reading this book. It is meant for those who want to expand their thinking about what they are hearing and learning about the new science discoveries and how this is reflected in the established theology of humanity.

Christian Pedersen

PART ONE

ALL THE THEOLOGY OF HISTORY, SIMPLIFIED

Theism is not simple and should not be simplified!

Stop anywhere it would be prudent to ask any question.

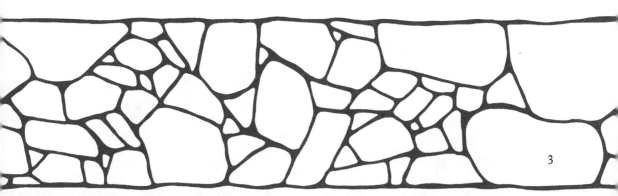

CHAPTER 1
Simple Overview of the World's Major Religions

There is nothing simple about religion. Religion has been a building block of society and culture throughout history. Brilliant thinkers have exercised their opinions and given direction throughout this time. This discussion has been creative, constructive, and sometimes dogmatic. Most of the discussion has been to the benefit of *Homo sapiens,* and some to strengthen positions of power or to create new dogma. Some would call it myth.

In psychology, we humans are conditioned to the story. Not necessarily the factual story, just the repeated story. To believe and reaffirm the story, we need to suffer or put extra effort into the story, and this solidifies our minds to the story. The story does not have to be correct—just good, believable, and cause us to put something of an effort into the storyline. Reflect upon some of the great stories of history. The French Revolution, communism, and German fascism are great examples. So, to understand the major religions of the world, we need to understand their stories. If we dissect the story, it is vital that we put it back together, or we lose the meaning of it.

The great creative minds of history have invented magnificent words and language to describe their stories, and so we must also decipher some of this language and its meaning. The language gives impetus and authenticity of authorship to each of the stories. So, for a while, at the beginning of this story that I am writing, I try to discuss the words and descriptions of the major religions of this world. I will attempt to use the words of the proponents of the various practices. I will inject some of my own understanding as well, which gives you the opportunity to think about the accuracy of my statements. Invoking you to think about statements is one of the major criteria of this book. Hopefully, it will also make this more readable and more unlike study notes.

We start by simplifying that which is not simple, and probably was never intended to be simplified.

Theism and Theology

Theological ministers will admonish anyone using these terms interchangeably. There is a slight difference in meaning. Merriam-Webster defines theism as:

> belief in the existence of a god or gods; *specifically*, belief in the existence of one God viewed as the creative source of the human race and the world who transcends yet is immanent in the world

In contrast to this, Merriam-Webster defines theology as:

> the study of religious faith, practice, and experience; especially the study of God and of God's relation to the world

Monotheism

Monotheism is the belief in a single God. This is the fundamental belief of Christian, Jewish, Islamic, and Sikh religions: a single creator God, with power over all things; thus, described as omnipotent.

Judaism and Christianity

The Jewish word for God is "Yahweh" and is formed from the word for "I am". I listened to a sermon describing how to pronounce this: a heavy breath inhaled, followed by a heavy breath exhaled sounding like Yahweh without any vowels in the word. Thus, the image of God is the very breath you inhale and exhale. The following text quoted from the Christian New International Version (NIV) Bible is from the Old Testament, which also represents the first five books of the Hebrew bible, the Torah. In these verses—said to have taken place between 2000 and 1000 BCE— God appears to Moses, who asks, who are you? God says, just call me the "I Am".

> [13] Moses said to God, "Suppose I go to the Israelites and say to them, 'The God of your fathers has sent me to you,' and they ask me, 'What is his name?' Then what shall I tell them?"

[14] God said to Moses, "I am who I am. This is what you are to say to the Israelites: 'I am has sent me to you.'"

[15] God also said to Moses, "Say to the Israelites, 'The Lord, the God of your fathers—the God of Abraham, the God of Isaac, and the God of Jacob—has sent me to you.'

"This is my name forever, the name you shall call me from generation to generation." (Exodus 3:13–15 NIV)

These passages also mention Abraham, Isaac, and Jacob. Abraham is the father, and the others are his son and grandson. They also are part of the story, as Abraham is said to have lived from 1986 BCE to 1859 BCE, predating Moses. In Genesis 22:1–19, we learn that Abraham was a very devout man who heard God ask him to sacrifice his son Isaac as an act of faith. Abraham went where he was asked, and just as he was about to sacrifice his son, an angel gave him a sheep to kill instead. God said he was pleased with Abraham for his belief leading to the potential sacrifice of his son in the name of God. This son, Isaac, along with his son, Jacob, became the leaders of the Jews. In addition, and relevant, in Genesis 16, Abraham's wife Sarah suggested that she was too old to conceive and so asked Abraham to have a son with her servant, Hagar. This son was called Ishmael, and he became the leader of the Islamic faith. Therefore, there is a connection between the Jewish faith and Islam, as well as Christianity.

The Jewish people believe that Judaism is the faith that God appointed them to have and be the example of holiness to the world: God's chosen people. The chosen people concept is ever present in the mindset of Jewish people and has also been the bane of their existence. Think Nazism and all the horrors Jews faced during that period of history. But this mindset was present long before the result of German eugenics. Jewish people have cloistered themselves through this concept for generations, frequently with racist results from those around them.

Many modern Jewish people may be non-theistic. They are Jewish by DNA, but do not practice their religious background.[2]

---

[2] Different types of Jews. https://www.answers.com/ Judaism has only two major sects. However, they have very different perceptions of each other, and these are the Torah Jews and the Liberal Jews. The main divisions between these sects is the ability to use non-Judaic source content to abrogate and modernize Jewish teaching

The largest three Jewish religious movements are Orthodox Judaism, Conservative Judaism, and Reform Judaism. The interpretation of Jewish law defines the differences between these groups: first the authority or power of the rabbis, and second the fundamental understanding of the State of Israel, as stated by the government of Israel.

Of course, the Jewish God, Yahweh, is also the precursor to the Christian interpretation of God, since the "Jesus of Nazareth" movement within the Jewish religion started with the teachings of Rabbi Jesus of Nazareth. Jesus is quoted several times as having said he was the "I AM", which did not endear him to the other rabbis within the Jewish faith because of the association with Moses. People following Jesus started "The Way", a movement extoling the values of this Jesus. This movement existed until about 300 AD, whereupon, courtesy of the Emperor Constantine, it became known as the "Roman" or "Nicene" Christian movement—independent of the Jewish faith but evolved from it.

The modification of Judaism by Jesus, pursued by the Apostle Paul after the death of Jesus, really accelerated in the year 70 AD, forty years after his death. That was when the son of the Emperor of Rome, Titus, entered Jerusalem and laid it to waste, killing almost all the leaders of the Jewish faith. This act of putting down the rebellious Jewish population coincidentally included the followers of Rabbi Jesus.

Military solutions were (and still are) brutal and indiscriminate. Following this military intervention, most of the human remnants or Jesus converts to Christianity were not Jewish by birth. Rather, they were those outside of Jerusalem, in other countries like Greece, and were only too happy to avoid Jewish practices like circumcision. Until that time, the followers of the "Jesus movement" were a subdivision of Judaism.

This Jewish Jesus branch was also the only faith or belief system, including pagan arts, that was evangelical, meaning they sought out and made welcome all new members. This included non-Jews and even Samaritans into their ranks, much to the distaste of all the other branches of Judaism. These, non-Jews who remained within the Jesus branch, now having inherited the Jesus movement or The Way, took exception to some of the tenants of traditional Judaism—things like adult circumcision as a God-mandated priority to believing and a tenant of belief in Judaism.

---

and practice. Torah Jews believe that such things represent a deterioration of Jewish identity and purpose, whereas Liberal Jews see Jewish Identity to be more internal or ethnic and the religious aspect to be secondary.

Do you remember my earlier statement about the story requiring sacrifice?

Women were treated much more beneficially in the Jesus movement than in any other belief systems existing at the time, and because of the persecution of the masses, there were a lot of single women, with children, who sought out friendly and safe relationships. They were supported with care, food, and assistance with raising children and their own health needs. This stood in contrast to the rest of society at that time, where widows were pretty much abandoned to the support of family or the generosity of a new husband. The status of single women and mothers was dismal in those days, and they had no status within the law. The story of these early Christians is described in the Bible within the entire book of Acts.

Most followers of the Jesus movement were from other areas of the world, like Greece. These followers in The Way, who continued to be severely persecuted, decided to form collectives for personal support. From that time on, the Jesus movement was evolving toward the Christian faith evident today. Of course, it was aided and abetted in 315 AD by Constantine, the Emperor of Rome, who through a vision became a proponent of Christianity. He also understood that the slow decay of Rome could possibly be alleviated by taking on the culture of an evangelical acceptance like the Christian faith. In his making Christianity the state faith, Christianity suddenly had structure, power, armies to defend it, and powerful people who were now at the top of the hierarchy.

The Christian faith is difficult to understand as a monotheistic faith since there is the belief that God the Father, God the Son (Jesus of Nazareth), and God the Holy Spirit are not three Gods, but one: a trinity. This becomes even more complicated when studying the Christian faith since you cannot find the Trinity in the Bible. It appeared after the Roman Empire adopted the faith. At the same time, the Roman Christian Church recognized the authority of the emperor, the pope, the saints, and Mary the mother of Jesus as deities (i.e., God-like and or having divine status by being a supreme being or creator of everything; appointed by God or anointed by God). Politically, this paired well with the politics of the day, avoiding paganism whereby the rulers were gods unto themselves, yet gave them the unquestionable authority of God.

Islam

Islam refers to God as "Allah". Allah, in Islamic theology, is the all-powerful and all-knowing creator, sustainer, potentate, and judge of everything in existence. Islam emphasizes that God is One, also all merciful and omnipotent. In Arabic, gender is either male or female, and "One" is neither.

Many people are confused about the difference between the words "Islam" and "Muslim". Islam is the religious theological practice of those who practice the Islamic religion. Muslim is the word used to describe the person who practices that religion.

Although Islam is claimed to be descended from Abraham through his son Ishmael, the same Abraham who was the precursor of the Jewish religion, it really came unto itself in 600 AD with the life of Muhammad.[3] Muhammad was born to wealthy parents in 571 AD in Mecca. He grew up in the vicinity and under the influence of Christian and Jewish people. He migrated to Medina, where he created/modified the religion of Islam. He died in 632. He is said to have been visited by God and therefore was a prophet of God.

Muhammad left oral and written ruminations of his visions, which were later transferred by scholars to written form and became the Quran. His life was also documented in books called the Hadeeth and the Sunnah. Muhammad also created the Islamic doctrine of Sharia law. In Islam, although the Quran is translated into all languages, it is considered absolute and final, with no modifications of interpretation possible. If questions of interpretation exist, the original text in Arabic is overriding. The Hadeeth and Sunnah are similarly considered resolute in interpretation. Sharia law is somewhat flexible and is interpreted by the various sects of Islam according to their modified beliefs.

Schools of Thought Within Islam

The largest body of believers is the Sunni, who represent more than eighty percent of believers. Second is the Shia, who represent approximately fifteen percent. Like all

---

[3] Abraham in Islam https://en.wikipedia.org/wiki/Abraham in Islam Muslims believe that the prophet Abraham became the leader of the righteous in his time, and that it was through him that Arabs, Romans, and Israelites came. Abraham, in the belief of Islam, was instrumental in cleansing the world of idolatry at the time. Paganism was cleared out by Abraham in both the Arabian Peninsula and Canaan. He spiritually purified both places as well as physically sanctifying the houses of worship. Abraham and Ishmael further established the rites of pilgrimage, or *Ḥajj*, which are still followed by Muslims today.

other religions over fourteen centuries old, this religion has numerous sub-branches that have formed over time. Each has different interpretations of Sharia law and variances of the fundamental methods of practice of Islam. There are numerous sub-movements that believe that Islam is not only a religion, but also a political system that should emphasize governance—the legal, economic, and social nature of a nation. This attests to the modification of Islam over the ages, just like in the Jewish and Christian faiths.

There are four schools of thought or movements—Sufi, Quranism, Ibadism, and Ahmadiyya—in Islam that are more social in nature, and which suffer lack of recognition within most of the other practices of Islam. Apparently, in recent times, many movements for the modernization of Islam have been started, and there are even movements for Islamic feminism, but change takes time, and results are slow to be perceived.

Currently, there appears to be a struggle of regression and progression within the Islamic faith, polarizing those within into acts of abhorrent behavior and, opposite to that, acts of generosity and caring. Incidentally, the country with the most Muslim believers is Indonesia, second is the Islamic Republic of Pakistan, and third is India. To visualize this, look at the governance of the country representing the highest percentage of its population as Islamic. This is Indonesia, which has varied from a nearly totalitarian regime to one of more tolerance in recent years and can give perspective to us today on the popular perception of the struggle within this belief system. Indonesia seems to have moved from an intolerant society of those outside their religion to one of more tolerance for a cosmopolitan society.

Sikhism

I live in an area that includes more than three hundred thousand people who identify as members of the Sikh faith. Knowing they are open to visitation; it is easy to attend their temples. One of the basic tenets of their faith is that they accept other faiths and welcome visits to their places of worship. However, also understand that Sikhs fiercely protect their temples and their properties, if threatened.

Just like the Jewish and Christian groups, there are numerous Sikhs who never grace the inside of the temple. They identify with the cultural term Sikh, but do not openly practice the religion. Their home region is the Punjab region of India and Pakistan.

Living in the Punjab in the 1500s meant being overrun by Hindus, then being recaptured by Islamic tribes, and then back again. The first Guru, "Nanak" Dev Ji (1469–1538), established the Sikh religion, which nearly coincides with the reformation of the Christian Church started in 1517. He developed the Sikh practice as a compromise between Hindu and Islamic traditions to hopefully live more peacefully in this war-like region. Part of his postulations said he was neither Muslim nor Hindu and referred to both as equals. That also means that he shunned the caste system of the Hindu.

The word Sikh means student. Sikhs signify God with the name Waheguru and meditate on this. Sikhism is characterized by several beliefs about God, as quoted below:

Sikhs believe that God:

- **Cannot be described and is neither male nor female**
- **Is eternal truth, timeless, beyond the cycle of birth and death, and self-existent**
- **Is both** *sargun* **(immanent – everywhere and in everything) and** *nirgun* **(transcendent – above and beyond creation)**
- **Created the world for people to use and enjoy**
- **Created people and made them know the difference between right and wrong**
- **Is present in everyone's soul but can only be seen by those whom he blesses**
- **Is personal and available to everyone**
- Is the only one to be worshipped; no images of God are to be worshipped
- Is made known by the grace of the Guru (GCSE Bitesize 2014)

Guru literally means "the one who dispels the darkness and takes one towards light" (Wikipedia 2018). Guru means teacher and all the attributes of being a teacher like molding others in their values, imparting knowledge, and inspiring and demonstrating a good example of their spiritual life.

In 1699, Guru Gobind Singh required that all Sikh members always wear five articles of faith. These are

1. Uncut hair
2. A special kind of underwear
3. A hair comb

4. A bracelet

5. The *kirpan*

The *kirpan* is a long-curved knife. Devout Sikh members still wear a replica of this under (or some wear it over) their clothing, if possible, and if you ask them about it, they will tell you it is to protect their personal "airspace". Airspace was defined by a spokesman for the Sikh temple during my visit there as the space they have around themselves. Much like the Western concept of "personal space", this is their personal "airspace" and it is to be protected, he said.

This statement would also subliminally reference their rationale for fiercely protecting their property and their religion, as demonstrated by Sikhs' past political activity. Some of this activity has been unacceptable to most of the world's population, as it has included blowing up airliners and murder of those in opposition to the Sikh way of thinking.

To me, both this statement and my own observations would indicate there are several degrees of liberal to fundamentalist types of Sikhism, just as there are in the Jewish, Islamic, and Christian traditions. If you live in a Sikh region, you will be familiar with the constant conflict within their religion about whether to sit on the floor or on a chair while in the temple. Not unlike the Islamic traditions of their Muslim followers, many Westernized Sikhs now choose not to wear saris or the cloth turban over their heads, and there is some chastising of them within their ranks by others within, who are not quite ready to drop tradition and integrate into their new adopted Western society.

In total, there have been ten gurus building tradition in the Sikh faith over hundreds of years.

Deism

Deism is belief in a supreme entity; an entity that cannot be described, is not seen, and obviously unable to be touched; a God that created nature and all-natural laws, under which our universe operates. This God does not operate in our daily lives and does not supervise events within our realm. God is perfect, and God created the laws of nature, which are perfect and therefore do not need the attention of God. Reasoning and human logic replace the faith necessary within a deist belief system.

This perfect God, who created perfect rules that cannot be broken, is a perfect place for those who have a science or math background, or for those who are in this perfectionist mindset. Although scientific law is constantly changing as we learn more and more about our universe, our psychological story of science gives the impression that scientific law is finite, unbreakable, and lasting. You cannot change the speed of gravity, for instance. Someone immersed in scientific study would see the appeal of this type of thinking.

Remember the definition of the human story in the introduction of this chapter?

What Deism does not do is nurture or forgive or love. You cannot pray to a God in deist thinking because you are wasting your time. Asking God to change a situation through prayer would be asking God to break the very laws that God created. It would not be realistic.[4]

Multitheism

Continuing in the theme of more than one God, Multitheism is the belief in more than one God. Hindu and the offshoots of it—Buddhism and Jainism—are the modern-day examples. Buddhism is non-theist (with some exceptions) but is included here as it is also originally derived from the Hindu faith.[5] Jainism is another. Scholars who study the history and culture of man, as it developed through the ages, will often state that Multitheism has the least propensity for war-like activity. Their reasoning is that if the religion supports both good and bad gods, the people of that system understand differences of opinion in a more tolerant way.

Hinduism

A very simplistic way to initially understand the Hindu religion is to first think about God the creator, God the observer or maintainer, and God the destroyer. Following this are several other deities who are also worshiped but with less reverence according to one tour guide I had in India.

---

[4] Most of these examples were read and paraphrased from "Deism", *All About Philosophy*, (Peyton, CO: AllAboutGOD.com, 2018) https://www.allaboutphilosophy.org/deism.htm

[5] From the Existential Buddhist. The claim of Non-theism is true in the sense that there is no God in Buddhism who is a Creator, Judge, or Deity-in-Charge. In Buddhist cosmology, the universe has always just existed and is continually evolving and devolving based on causes and conditions. https://www.existentialbuddhist.com

God the creator is Brahma, God the observer is Vishnu, and God the destroyer is Shiva. Here is a quote from an official representative of the Hindu religion.

> *Hindus acknowledge that at the most fundamental level, God is the One without a second—the absolute, formless, and only Reality known as Brahman, the Supreme, Universal Soul. Brahman is the universe and everything in it. Brahman has no form and no limits. Brahman is Reality and Truth.*
>
> *Thus, Hinduism is a pantheistic religion: It equates God with the universe. Yet Hindu religion is also polytheistic: It is populated with myriad gods and goddesses who personify aspects of the one true God, allowing individuals an infinite number of ways to worship based on family tradition, community, and regional practices, and other considerations.*[6]

(Srinivasan 2011)

Hinduism includes, among others, the following gods and goddesses in addition to Brahma, Vishnu, and Shiva: Ganapati, Rama, Krishna, Saraswathi, Lakshmi, Durga Devi, Indra, Surya, Agni, and Hanuman. Also, while in India I noticed many Hindus worshiping at the temple for Hare Krishna. The Hindu movement began with the arrival of the Aryans 1000 BCE. The Aryans also brought the caste system into being as a way of controlling the local population, which was much larger in number than the invading Aryans.

The caste system divides Hindus into four main categories: Brahmins, who are the priests, scholars, and teachers; Kshatriyas, who are rulers, warriors, and administrators; Vaishyas, who are farmers and traders; and Shudras, who are the untouchables or laborers and service providers. Each member of their own caste is to reconcile or celebrate their status and retains that status for life, even for the lives of their offspring.

Another facet of Hindu daily life is a complete belief of astronomy ruling their daily lives. Nothing is done without first consulting the astronomical pages of the stars. Interestingly astronomy was created by the Babylonians approximately 1790 BCE.

---

[6] Hindu Gods and Goddesses By Amrutur V. Srinivasan Part of Hinduism For Dummies Cheat Sheet

Buddhism

Buddhists believe in humanity. However, their roots are from the Hindu tradition. The Buddha was born in Nepal, 500 years before Christ. At the time, this area was entirely Hindu in practice and had never contacted the Western religion of Judaism. The Buddha also was a witness to mendicants,[7] peoples who left their positions in life and constantly meditated. This perhaps is the reason that the Buddhist tradition is full of meditative practice. I also understand that the Roman Catholic Church accepts the practice of meditation, as prescribed by Buddhist monks, since it is considered a practice, not a religious endeavor.

The goal of a Buddhist is enlightenment. At the time of the Buddha, it was understood that enlightenment could only be attained by leaving all your worldly goods behind, including your family, and entering a convent with the intent of practicing meditative dogma until enlightenment was attained. It wasn't until about 200 AD that this practice changed somewhat, and enlightenment was possible without all the self-sacrifice.

I have been to several Buddhist countries and spent a brief time with monks. There are a few people within those countries who not only meditate but also believe in "the Buddha" as a deity and are openly rejected by most monks for this.

Although the Buddha himself was a theist, his teachings are non-theistic.

Mendicants

The term "mendicant" is used in the description of Buddhism, so let's include a description of mendicants here also. Mendicant is used to denote a holy person who is committed to abstinence from sensual pleasures. This could include members of religious orders or persons who worship their own individual concept of religion or practice. Eastern mendicants[8] were influential with the Buddha in deriving their practice of meditation. These mendicants meditated as an everyday act of life. These Eastern mendicants are often confused by students of theology with the term "Western mendicant".

---

[7] A beggar

[8] While mendicants are the original type of monks in Buddhism and have a long history in Indian Hinduism and the countries that adapted Indian religious traditions, they did not become widespread in Christianity until the High Middle Ages. https://en.wikipedia.org/

Western mendicant orders are understood to be Christian religious orders that adopt a lifestyle of poverty, traveling, and residing or resting in populated areas for purposes of preaching, evangelizing, and ministering to the poor. The description is a little like being a Gypsy, but for religious reasons. Some mendicant sects live in a single stabilized community exclusive to themselves, where members work, and property is jointly owned by all, including land, buildings, and other wealth; a commune. But this was not the norm. They normally do not own property, shun work, and embrace a poor and traveling lifestyle. Mostly, for sustenance they depend on the goodwill of the people to whom they preach.

Jainism

Like Buddhism, Jainism is an offset of the Hindu faith. As with Hinduism and Buddhism, Jainism is the third ancient Indian religion, and it is still in practice. Although like and using many of the same practices of Hinduism and Buddhism, it is separate from them.

The name Jainism means "to conquer". It is a reference to the battle that the monks and nuns fight against worldly possessions and self-possessed needs. All this is to gain enlightenment, or all-knowing (omniscience), and have purity of soul. Individuals who have gained enlightenment are called Jina within the tradition. Jina means conqueror. Worshipers who have not yet gained enlightenment are called Jain, meaning follower of the conquerors, or Jaina.

The practice of Jainism has mostly been confined to India. With the large influx of people from India to English Commonwealth countries and the United States, the practice of Jainism has also spread to these countries.

I understand that within this practice the followers will not eat anything that has to be killed and so are vegetarian. Apparently, there are some who so vehemently oppose killing that they will not even swat a mosquito when it is biting them, or who will step around a spider in their path.

Dualism or Gnosticism

The term dualism has its roots in history in the form of Gnosticism. Gnosticism was perceived to be a much-reviled form of worship that existed only in the first three hundred years of the Christian Church.

Gnostics were perceived to be perhaps the most dangerous heresy that threatened the early Christian Church.[9] Early philosophers, including Plato, created a premise based on two false notions. First is the dualism of spirit and matter in conjunction. Gnostics believe that matter is evil, and spirit is good. Because of this premise, Gnostics believe anything done in or by the body, even the grossest sin, has no meaning because reality is only in the spirit realm. Second, Gnostics claim to possess superior knowledge, unknown to the rest of humanity.

Gnosticism is derived from the Greek word *gnosis*, which means "to know". Gnostics claim to have this superiority, not from the Bible, but acquired on some mystical higher plane of existence. Gnostics see themselves as a privileged class, elevated above everybody else by their higher, deeper knowledge of God, among other things.

How does that fit with your definition of narcissism? Maybe a touch of ego, too?

Panentheism and Pantheism

These are terms regularly in use with discussion of theology by theologians. Let's explore what they mean since they are not in common use.

Panentheism, literally "all in God", is the statement that God is present in every aspect of the universe and yet is bigger than time and space; a fair and just God who comes down to forgive and to love us and all living things.

Pantheism holds that God and the universe are identical, that God is out "there", not present in us or to us in every way. This is much like the discussion on deism where the perfect God who made everything is out there, but not involved in our everyday lives.

Much Hindu thought and Buddhist philosophy are highly characterized by panentheism and pantheism.

Atheism

Merriam-Webster defines atheism as: "A lack of belief or a strong disbelief in the existence of a god or any gods."

There are those who would use this definition to categorize atheists as either negative atheists (agnostics) or positive atheists.[10] They would say, agnostics, while

---

[9] https://www.gotquestions.org/Christian-gnosticism.html

[10] https://www.urbandictionary.com/define.php?term=Atheist

they don't believe in a god, do not positively assert that no gods exist or are on the fence. Positive atheists, however, admonish that there are no gods.

North American Native Spirituality

At this juncture, we must examine the indigenous culture and spirituality of North American natives. This includes the Inuit, or Eskimo, in the far north to the grass plains Indians in the south. This discussion is restricted to the North American indigenous peoples whom I am just beginning to understand in a limited way. Other native populations on other continents, and especially their spiritual background(s), are beyond my focus and current knowledge.

I refer you to an excellent book, *Indigenous Healing*, by Rupert Ross, a lawyer and Crown Attorney in Ontario, Canada. (Ross 2014) He tries to explain the heretofore unexplainable with regards to Native thinking, culture, and spirituality. His comments are remarkable and revealing about what we have completely missed in our integrating of Native culture into North American culture. Under the chapter on eugenics in this book, I will also explore how we mishandled our historical integration of Native peoples into our culture.

Spirituality is the only way for me to try to explain the religion of Native culture. I use the word "spirituality" in a loose and condescending way in other circumstances within this book. Spiritualism was the term used by traveling magicians and escape artists, like Harry Houdini, around 1900. Please understand that the meaning of spirituality here is far from demeaning and condescending; it is the only way I can describe what I have come to understand of Native religion, and I still have much to learn about this.

Listening to those Natives who have become friends, I understand they regard everything as relational. Since they crossed the ice bridge from Asia, they have understood who and how they are in relation to everything else in our complete world. Relationships are the basis of their everyday life and belief system.

When I was much younger, I was working in the high Arctic. Yellowknife was considered my home base, but I hardly ever came that far south. One day, on a rare day off, and walking the street in Yellowknife, I came across a very young Inuit woman wearing her Native seal-skin parka and carrying her baby inside her hood, on her back. This was my first encounter with an Inuit who appeared to be away from her remote home for the first time, still in her Native apparel and seemingly as innocent as a lamb. I was overcome with the sincere simplicity she represented in the most genuine way. As I walked past, I made a non-confrontational complementary remark in recognition of her presence. This very young woman, who I believed had never experienced civilization as we know it, stopped, smiled in the most charming way, and said thank you to me. I felt then, and still do to this day, that she saw me in a completely different way from that which I had experienced my whole life. It seemed as if I was someone who momentarily crossed her spiritual path in life and instantly appreciated the connectedness of a momentary interaction. I have never forgotten the simplicity and spiritual beauty of it.

We are being told through the government and Native reconciliation process of today that Native spirituality is always relational. They are part of the whole, part of the process by which the Earth provides life, using everything upon it to modify and support everything else. *Homo sapiens*, as they understand it, are not the center of the whole, just a part of the process, which must be honored, recognized, and worshiped. Everything has value in relation to everything else—a very different perspective from the typical Euro-centric notion that *Homo sapiens* are the center of everything, the alpha male of all living things.

In his book, *Indigenous Healing*, Rupert Ross provides a story about the Native perspective of relationship between species. He described an oft-repeated Native story of wolves being removed from Yellowstone National Park by allowing them to be hunted until extinct, and the consequences arising from that.

Briefly, by allowing the wolves to be hunted to extinction in the park, the elk population that the wolves feasted upon exploded. As it did, their food supplies were put under heavy pressure. The elk started eating the leaves of the bushes along streambeds. Over time, those bushes died out, which meant their root systems also died out. The soils that were once held in place washed into the streambeds. As the water clouded, the fish habitat was altered until certain fish stopped spawning. Some species of fish were then faced with extinction. Without the wolves, the park became a much different and poorer place. Once the wolves were reintroduced, everything began to reestablish and reassert their roles, and with surprising speed. The elk population was reduced, the bushes grew, soil was retained, water regained clarity, and the fish spawning was restored.

The object of the story is that wolves are relational to the whole system. This appears to be the purest form of Native spirituality and the only way I can express it.

Stop, Think, Discuss

I offer this thought: Knowledge, theology, and the application of reason are of little value, if not utilized wisely. They can be manipulated for nefarious intentions.

As you read throughout this book, think about how knowledge and wisdom, scientific or theological, are negatively used for the benefit of those who would create, supplement, or enrich their power base or for financially rewarding themselves. Can you name a few instances?

Question one: Reviewing all these definitions, what is the divine intention of God and what is the hand of man? Is God even involved? Is all this God's creation or man's?

Question two: What is the *ontology* of these statements? That is, what is the actual information, free from our personal bias, of these statements? And what is the *epistemology*? That is, what is the knowledge, as learned by us, including our preconceived notions, of these statements?

Question three: In all the definitions, what appear as ideas created by God? Which were created by man, if any?

Question four: Everything that happens almost always has a precursor that makes it happen. Discuss your ideas of what made all these theologies happen. How did all the scholarly depth of thought within all these religions come to the mind of *Homo sapiens*?

Question five: How would you describe these statements of theological simplification in your own words?

Question six: Are there issues in these statements with which you disagree? Has Harry Potter—creative thinking—taken over the dialog on this issue?

CHAPTER 2

The Concept of the Devil

Judaism, Christianity, and Islam view Satan as a fallen angel. A fallen angel is an angel that has lost favor with God and has been cast out of heaven. These deities, now expelled, have gone to the spirit world to the opposite of goodness, love, and forgiveness, but still have influence upon us, the residents of this world. Even though they rebelled against God, they are under God's control. Therefore, Satan with god-like powers, "masquerades as an angel of light" (2 Cor 11:14 NIV), deceiving humans just as he deceived Eve as a serpent in the beginning (Gen 3 NIV). Satan is mentioned throughout the Christian Bible in both the Old and the New Testament, but always in connection with Christ.

Muslims believe that the devil does indeed exist, but Islamic belief is different to that of other religions. Muslims, who are the followers of Islam, believe that the devil, named Iblis, is from the Jinn. In Islam, Jinn are a kind of spirit creature and can manifest evil by rejecting Islam or manifest good by accepting it. Jinn have free will and can oppress and possess human beings, animals, and objects. So, the devil is not a fallen angel as the Judeo-Christian beliefs describe him. According to Islam, Iblis was from among the righteous Jinn before his arrogance and defiance to God caused his downfall. Iblis refused to bow down to Adam when God created Adam. This was because he, Iblis, stated he was created from fire and therefore superior to Adam, who was created from mud. He then swore to misguide as many people as possible and declared war against those who try to follow God's religion.

Sikhs do not believe in a devil, or Satan, but rather believe that the concept of demons or devils is just a wrestling contest with one's own inner ego. Sikhism teaches that ego is the prime cause of evil doing in humans. Sikhs have a very good grasp of ego and teach that human ego has five basic components—pride, lust, greed, attachment, and anger—all of which lead to evil doing.

Buddhism has a plethora of gods and each has a personality. Some are good, some are evil, and Buddhists describe an escape from evil in their teachings. They speak philosophically of Mara the Evil One, who is the personification of evil, or the devil. Mara represents temptation, sin, and death and is identified with Namuche, a wicked demon in Indian mythology and with whom Indra, the god of thunderstorms, struggles. Namuche is mischievous and prevents rain, which creates drought. The struggle between Namuche and Indra, in which Indra forces him to give up the rain, restores the Earth with life.

Hinduism has a great deal in common with Buddhism. There is no single god as Satan simply because all the gods have personalities with good and bad sides. Therefore, there is no need for a single god to monopolize suffering. Whatever god's personality is in control of their faculties in their activity is good and, when out of control, destruction or evil exists. Therefore, if your mind is clear and you have control, it is good; you benefit society. If you are not in control, you misuse your power, which makes you evil. As described, good and evil reside within us, and it is up to you and how you use your mind that determines a state of good or evil. The universal truth of Hinduism is that everything wants to be in equilibrium and your state of mind is your state of god. Hindus further state that good or evil resides within us but never in exact opposition to each other. Therefore, good and evil are never in equilibrium within us.

Stop, Think, Discuss

I would say that a modern way of describing Satan would be to describe evil as "an outrageous act of such malevolency as to invoke the complete revulsion of the majority of upright-walking and thinking citizens inhabiting this lonely orb."

What would be your description of Satan? Can you see evil in the world? Have you seen evil? Explain.

Question one: In these few statements about evil and Satan, what is your understanding of the divine intention of God and what in them is made up by the hand of man? Do you even see intention or acts of God involved in our descriptions?

Question two: What is the *ontology* of these statements? That is, what is the actual information, free from our personal bias, of these statements? And what is the *epistemology*? That is, what is the knowledge, as learned by us, including our preconceived notions, of these statements?

Question three: What appears to come first in this? Was it an idea created by God or by man? Is God involved at all? Is there evil, as represented by these theologies?

Question four: In our statements of evil and Satan, do you see anything that would be necessary for the storyteller to tell to make man accept this as reinforcement of the story of the existence of God? That is, is the story of evil a storyteller's supplication?

Question five: How would you describe the analysis of this chapter in your own words?

Question six: Are there issues in these statements with which you disagree? Describe.

CHAPTER 3

The Concept of Heaven and Hell

Christian Belief in Heaven and Hell

Christians have a belief in heaven, even though there is no specific statement on this in the Christian Bible, other than the resurrection of the body, and the metaphorical interpretation of Bible passages, especially throughout the New Testament. There is the clear description of Jesus rising to Heaven, but not anything that directly describes *Homo sapiens* as doing so. Only metaphorical interpretation applies to us humans. A place where God dwells with all nations on Earth, where sins have been forgiven, where life is glorious, fulfilling, and everlasting. A place where you are in the presence of God and revel in the perfection of God. Some Christians think of heaven as only symbolic and allegorical. But then Christianity has become a widely interpreted theology, from fundamentalist to allegorical. One thought is that Christianity will renew here on Earth in a glorious state of heaven on Earth.[11] This idea of a place where God comes down to us on this earth, which has been restored and is now heaven, is a very different concept from what is found in other religions.

Many Christians also believe in hell. The existence of hell is a debatable point of view for some scholars, as they would propose that Dante wrote of the nine circles of hell in his poem, *Inferno*, around the year 1300, and his perspective became the populist viewpoint of Christian belief; a point of view that is not supported by the Christian bible, as well as the concept that hell is not a Christian concept of forgiveness and eternal love where someone is dammed to Hell for life after some misdeed.[12]

---

[11] This is not a predominant concept and should not be confused with mainstream Christian dogma.

[12] "What if the muting of hell is due neither to emotional weakness nor loss of Gospel commitment?" writes Edward Fudge, whose 1982 book, *The Fire That Consumes*, is widely regarded as the scholarly work that jump-started the

Additionally, to prop up their finances, the Roman Catholic Church adopted a payment system from around 700 AD to the reformation in 1517 AD, whereby they controlled purgatory, the gateway to heaven and hell. Your payment would purchase indulgences to avoid the waiting time in purgatory. Originally, indulgences[13] were created by the early Christian Emperors of Rome for certain occasions like Easter. A pardon or a pass on penance for your recent sins, indulgences were originally a gifting of grace given as celebration to a joyous populace in the early Roman Christian Church as a day in your life when you wouldn't be punished by God for your actions. These indulgences copied what emperors had previously been doing prior to adopting Christian values. Then, from about 700 AD and especially aggressive around 1400 and into the 1500s, this became a way of raising capital for the overall financial health of the Roman Catholic Church. This was also the precursor to Martin Luther starting the Reformation. However, during the Reformation in the early 1500s, and much later in the 1850s, Christian dialog emphasized the following with regards to the concept of hell:

A proper interpretation of the Bible for the words in the Old Testament that use *sheol*, or hell, mean simply the state of the dead, a world that cannot be seen; there is nothing about good or bad, happy or sad. It is not described as a separate place of hell or a state of endless castigation. This is not revealed within the Jewish Law of Moses, either.

Within the New Testament, the word *Gehenna* is used twelve times and is misinterpreted as hell by some Christians.[14] This word is repeated in the gospels, which each in turn repeat the same stories in each chapter of the Bible of Matthew, Mark, and Luke with their individual written records of the life of Christ and reduce the use of the word *Gehenna* to only six or seven uses. *Gehenna* is meant as a word to

---

current debate. "What if the biblical foundations thought to endorse unending conscious torment are less secure than has been widely supposed?"

[13] Historians of the Roman Empire use the word in the technical sense of *remissio tributi* or *remissio poenae*, concessions that the emperors customarily made on certain occasions. It was also used to indicate *abolitio*, a sort of amnesty decreed on joyful public occasions. (Thus in the Carolingian era, *indulgentia* was still being used as the technical term for the remission of penalties or taxes). https://www.catholicculture.org/

[14] https://creationconcept.info/ In the New Testament, the word Gehenna occurs 12 times. These are in Matthew 5:22, 5:29, 5:30, 10:28, 18:9, 23:15, 23:33, Mark 9:43, 9:45, 9:47, Luke 12:5, and James 3:6. The idea of infernal torment of unbelievers was introduced in ancient Rome as a means for controlling the people. We should look in the scriptures for the significance of Gehenna, rather than assume that human tradition and speculation is correct.

describe any severe chastisement, especially an infamous type of death—not hell, as depicted by some modern, perhaps more fundamentalist, Christians.

The Apostle Paul wrote vigorously during his vast travels across Europe and the Middle East. His Book of Acts describes the early years of the Christians prior to the Roman Empire adopting Christianity as the state religion. This was a period of great distress for those persecuted about their belief system, and Acts is very insightful of the core values of Christianity at that time. During those thirty years after the death of Christ, never once is "hell" mentioned by any apostle or proponent of The Way or that anyone will suffer the punishment of *Gehenna*, or hell, as it is now promoted, in even the most insignificant way. Dante has had more effect on our history of hell than the original Christians following Christ.

All these facts are presented to postulate that there is not the least hint of hell within the use of the word *Gehenna* in the old Christian Church, originally called The Way, and prior to the Roman Empire's adoption of it.

Jewish Beliefs of Heaven and Hell

Like other religious traditions, Judaism offers numerous views on the afterlife, including some like the idea of a heaven and hell that parallel those of Christian teachings. Traditional Jewish thinking about heaven and hell is extensively discussed by their spiritual leadership. In modern times, Jewish theologians have generally avoided the subject, instead using a modern biblical narrative that focuses on life on Earth.

Reading scraps off the Internet, we quickly come to understand that being a Jewish person means being one who has the DNA of a Jew. The Jewish religion is now secondary in life, much like Christianity, as observed by the lack of impetus of those within their respective religions. Along with that is the modern idea of the lack in belief of heaven and hell, just like those abandoning the primary role of Christianity in their life.[15]

The question of geography arises. Where is heaven and where is hell? Perhaps the kingdom of God is here on Earth? The Torah and the Talmudic rabbis wisely understood the dangers of describing heaven and hell. They could misinterpret the

---

[15] *The Jewish Journal*, June 09.2019. https://jewishjournal.com When it comes to the subject of the existence of heaven and hell, most contemporary Jews—meaning Jews who have graduated college, who are essentially secular, and who consider themselves progressive—know exactly where they stand: There is no heaven, and there is no hell.

intentions of heaven and hell because of the idea of reward or discipline in the after-life, for that is what heaven and hell are about. Heaven means there is reward after this life, and hell means there is punishment after this life.

Sikh Heaven and Hell

The Sikh religion discusses the existence of heavens and hells, but only as temporary places. A Sikh neither fears hell nor wishes for heaven. Anyone who follows the true Guru gives both heaven and hell no place in their thinking. Heaven is only a place for physical or sensual pleasures and, therefore, is insignificant. A Sikh only desire to live in the presence of God; a God recognized by them through their everyday actions.

"(Those who understand God's mystery,) they are exempt from incar-nation into heaven or hell."[16]

Sri Guru Granth Sahib (Ang 259)

A Sikh does not wish for any physical reward in death. The Sikh concept of "Sach Khand" (the Realm of Truth) is much different from the Islamic and Christian concepts of heaven or the Hindu concept of *swarg*. Because hell and heaven are not given importance or significance, some Sikh scholars have interpreted that there is no actual place as heaven or hell. (Juliette Bently 2016)

"What is hell, and what is heaven? The Saints reject them both. I have no obligation to either of them, by the Grace of my Guru."

Sri Guru Granth Sahib (Ang 969)

Heaven and Hell in Islam

Heaven is described as paradise in Islam with rivers of pure water, milk, and wine, even though alcohol is forbidden here on Earth. Salvation is by your works here on Earth, not by faith or the grace of God. There are similarities to some Christian the-ologies here, too, even though that would be disputed by most Christians. Muslims will be in gardens enjoying what the Lord has brought them, avoiding the fire and

[16] Sri Guru Granth Sahib. The Guru Granth Sahib is unique among the world's great scriptures. It is considered the Supreme Spiritual Authority and Head of the Sikh religion, rather than any living person. It is also the only scripture of it's kind which not only contains the works of it's own religious founders but also writings of people from other faiths.

drink with happiness because of what they have accomplished here on Earth. Their beds will be arranged in ranks according to their deeds, and they will be married to many lovely, fair women. You could ask here if this is heaven for Islamic men and hell for women, according to modern Western standards. Taken literally, this could be acid in the mouth of half of the population of the Earth. But reading the Torah and the Bible, we see discussion of slavery and women are not much different in tone from this.

Jahannam is what hell-fire in called in Islam, it is an afterlife of punishment. Punishment will be with eternal fire and pain. It is horror-movie frightening, dark, and unfriendly. It is blazing fire, with boiling water to drink, and poisonous food to eat. People will want to live again to correct their ways and begin to understand the reality of the afterlife. Islam teaches that heretics will spend an eternity in *Jahannam*, while believers who made errors during their lives will "taste" of the punishment but will ultimately be forgiven by Allah the Most Merciful.

Buddhist Concept of Heaven and Hell

The Buddhist heaven and hell are unlike those of other religions. There is no reward or punishment for an individual's deeds or misdeeds on Earth. Buddhist heaven and hell are temporary, a place where individuals are reborn according to the lives they lived on Earth. Once they have spent a predetermined amount of time in one of these two places, they are born again. Therefore, reincarnation is the central theme of this religious format.

Hindu Concept of Heaven and Hell

The Hindu concept of heaven is cosmological in origin. There are two levels, or planes, called Bhuva Loka and Swarga Loka, which together mean good kingdom, or heaven, as we would call it. This is a paradise of pleasure, where most of the Hindu Deva reside, along with the king of Devas (loosely interpreted as angels or supernatural beings), Indra (the king of the gods), and beatified mortals (a dead person who has been blessed).

Heaven and hell are temporary places where you go to do penance while you wait. Your faith is mathematically counted as to your deeds here on Earth. Good deeds give time in heaven, while bad gives time in hell. It is a formula, and you get to go both places before returning to make amends or continue a better path to enlightenment. It is only after liberation that a soul attains divine peace and divine companionship.

Hell is not eternal hell. It is more hellish states of mind and pitiful rebirths for those who think and act incorrectly. So, hell is a personality issue—fear, hate, jealousy, bigotry, and anger—brought about by one's own thoughts, actions, and deeds. Positive karma will get you out of hell, and therefore hell is only temporary.

Atheists' Heaven and Hell

Atheists pretty much do not believe in anything that requires faith to understand. The proofs of science and the discussion of philosophy are a part of their vocabulary, but theology is anathema to them. Therefore, there is no belief in heaven or hell. We exist, and we die. Dust to dust, with hours of searching for meaning in between. Coincidence is probably the closest an atheist will ever get to discovering the meaning of "why" in human endeavor.

Some atheists are enchanted by the science they see in their limited understanding of theology. They want proof to verify ideas, and science does that very well. Seeing into the future is impossible for anyone, never mind an atheist, and the concept of an afterlife, as proposed by a faith, does not compute.

For some atheists, especially in the Western world, their belief is not about heaven and hell, but rather about not being a Christian. They have derived this from abusive Christian practices around them over the years, and they have shunned this form of piety.

Stop, Think, Discuss

Think about Christian hell and the medieval Roman Catholic practice of paying indulgences to avoid staying in purgatory or hell as relief from punishment for our sins, and which, by the way, was one of the root causes of the Reformation.

Considering that indulgence program, could the modern-day government's new carbon tax payment, as relief from punishment for our environmental sins, be the precursor of a new modern Reformation compared to an indulgence payment of old?

How much has political bureaucracy changed in six hundred years?

Question one: In these statements, what is the divine intention of God and what is the hand of man? Is God even involved? Is this God's creation or man's? Do heaven and hell even exist?

Question two: What is the *ontology* of these statements? That is, what is the actual information, free from our personal bias, of these statements? And what is the *epistemology*? That is, what is the knowledge, as learned by us, including our preconceived notions, of these statements?

Question three: Were heaven and hell ideas created by God or by man? Compare your thoughts, religion by religion.

Question four: *Dante Alighieri* wrote a lengthy poem called *Inferno*, started in 1308 and completed in 1320. In addition to magnificent paintings depicting several levels of hell, Christian stories commenced being told of the rigors of hell. Did Dante have more influence on the concept of hell than Christian teachings?

Question five: How would you describe heaven and hell in your own words?

Question six: Are there issues in this chapter with which you disagree? Does this section of writing take too much license with the facts?

CHAPTER 4
Books, Laws, and Literature Supporting Theism

The Sacred Texts and Versions of Christianity

Christianity has morphed from the original days prior to Constantine to what today is a variance of opinion and fact that is difficult to follow or even understand. Certainly, the Roman Catholic tradition is around, but it is much modified in practice and even power structure. Offshoots—started by Luther and Calvin, creating the Protestant Christian Church in the early 1500s—have also evolved into branches of Christianity that would often cause one to pause while reflecting upon their dogma. How do these modified branches of Christianity relate to the original practice of 2000 years ago? This is not an exclusive revelation to Christianity either, but here we are breaking down the various prominent theologies and their literature, so we are discussing Christianity.

The Eastern Orthodox Catholics were likely the predominant Christian movement 2000 years ago in Russia, the Baltics, Hungary, Poland, the Ukraine, the Middle East, and so on. They became severely diminished when Muslims conquered their lands and forced them out around 1000 AD, thus causing the crusades. The Roman Catholics remained in the West, also mostly due to the Crusades between 1100 and 1200 AD, which held Muslims out of their region. After the Reformation in the 1500s, Protestant Christians became a prominent branch of Christianity, in large part due to the printing press, which published their popular criticism of the Roman Catholic practice at the time. Compare that to today and the revolution of the Internet. It is a similar phenomenon.

Protestants occupied the northern regions of Europe, the United Kingdom, and Switzerland, and eventually moved across North America with the colonization encouraged by persecution of some sects by some political states in Europe. The popular divisions of Protestant (reformed) Christianity are Anglicanism (or Episcopalians), Calvinism (in the forms of Continental Reformed Church, Presbyterianism, and Congregationalism), Lutheranism, and Methodism. Then, within the Protestant movement, branches of Christianity evolved that often become difficult for many of us to relate to the original Protestant concept. Statistics indicate that Christianity is the largest religious belief system in the world today with 2.2 billion followers. Statistics are for you to interpret because there could be a huge variation of opinion on whom statisticians are counting as Christians.

Regarding the rest of the world, there are Coptic Christians, mostly in Egypt, representing the ancient Christian Church of the Middle East, as well as Christians in Africa, Asia, and South America. Listing everyone and all their variances could occupy the pages of several volumes. All them are Protestant and, to some degree, Roman Catholic sects in the various countries have some variance of practice but are mostly known and understood by us citizens of the world. Let's focus on some of the lesser known internationally, yet more-known sects within North America for a minute.

Let's look at the first church reformed after the break with the Roman Catholic tradition. That version, now called Lutheran, was originally brought about by Martin Luther in 1517. Luther nailed ninety-five criticisms onto the door of a Roman Catholic Church in Germany, primarily but not exclusively regarding their fund-raising by selling indulgences or pieces of paper forgiving sins and reducing time in purgatory. (If you are interested in this, lookup the 95 Theses for reference.[17]) He also took it upon himself to translate the Holy Bible into the common language of the people, thus gaining tremendous support from the populace. After much debate, extreme criticism, and the possibility of being burned at the stake as a heretic, the Lutheran Church was founded. The small and large catechisms were authored by Luther himself and stand alone today as foundational in Lutheran theology. Other documents of Lutheran theology were written at nearly the same time by Philipp Melanchthon, a good friend and mentor of Luther. They are the Augsburg Confession, and the *Apology of the Augsburg Confession*. After Luther died in 1546, the *Book of Concord* was written by Jakob Andreae and Martin Chemnitz in 1584.

---

[17] What Are Some Facts About Martin Luther's 95 Theses? https://www.reference.com

This book was requested by rulers within the area occupied by Lutherans to avoid controversy and give clear direction in the theology of Lutheran literature. Probably the most basic tenet of Lutheran belief is the practice and use of the word "grace". This is found in the Bible in Ephesians 2:8 For it is by grace you have been saved, through faith—and this is not from yourselves, it is the gift of God— **9** not by works, so that no one can boast. (Eph 2:8–9 NIV)

Lutherans view grace like a Christmas present of forgiveness, freely given by a God—not earned by works or other interventions—who comes to you, listens to you, and is active in your everyday lives. As a forgiven member of the Lutheran ministry, you are expected to do good works, but not for recognition or keeping a tally record for your final days or to demonstrate that you are better or more religious than others who do not do good works. Rather, you are expected to do it because it is what a good Christian does and is expected to do. There are no bonus marks for good works when you pass into the kingdom of God. Many years have passed since the Reformation and so, like all other theologies that change with cultural changes and other ideas, so has the Lutheran Church branched off into other versions of the same theology, but with different emphasis on the implementation of their version of theology, including some who are literalists interpreting the Bible simplistically and without metaphor—for instance, believing that the world is only about 6,000 years old.

Just to describe a few other Christian sects, known but less common than mainline Christianity in North America, let's start with the Mennonites. They are first and foremost members of Anabaptist denominations named after Menno Simons (1496–1561) of Friesland, a province of the Netherlands. With his writings, Simons formally expressed the teachings of preceding Swiss founders. A partial introspection of Anabaptist writings reveals that a decision to follow Jesus Christ could only be made by people mature enough to be cognizant of sin. Their understanding of scripture in the Bible is that baptism is a public sign of their choice and their commitment. Thus, he went against the common practice of baptizing infants. Because of this, they were called "Anabaptist", meaning "re-baptizers". Incidentally, the Netherlands was originally composed of both Holland and Belgium, as we know them today. Religious conflict between Protestant and Roman Catholic theologies caused the country to separate into two, with Holland primarily Protestant and Belgium Catholic.

The Amish are included in the followers of Anabaptist teachings, but are slightly different from and yet very similar to Mennonites. Also include Hutterites among the Anabaptist followers. Anabaptist followers are often viewed as very fundamental because of their specific clothing and their lifestyle, which may include communal living, communal ownership of property, and aversion to alcohol, coffee, and dancing (depending upon the branch of Anabaptism you are looking at). However, within this same sect of Anabaptists are the modern Mennonite Brethren. They are indistinguishable within society in their appearance and behavior. They promote Jesus Christ, just like their fundamentalist relatives. They work hard, do not live communally, have great integrity, are pacifist by nature and by teachings, and are generous in ways that others in society would find unusual. Reading statistics would suggest that Anabaptist followers' number about two million in the world.

Much the same in tradition of only baptizing adults is the Baptist Church. However, they do not allege they are Anabaptist. First started in Amsterdam in 1609 by John Smythe, an English separatist, the Baptist Church is a very popular Protestant movement, particularly in North America, with fifteen million followers around the world.

Another popular branch of Christianity, which is often viewed as remote in belief and structure by other Christians, are the Mormons. They belong to The Church of Jesus Christ of Latter-day Saints. Their literature indicates that like original Christians, they believe in God as Father and his son, Jesus Christ, and the Holy Spirit. They also believe in modern prophets, like their founder, Joseph Smith. Mormons believe that returning to God requires following the example of Jesus Christ and accepting forgiveness through the practice of baptism. They believe that Christ's church was restored through Joseph Smith and is guided by living prophets and apostles. They use the Bible like other Christians but emphasize the Book of Mormon for interpretative use. Many shun caffeine and other humanistic pleasures as a reflection of living a proper life. There are sects within their Church that believe in polygamy, and even some extreme examples where they believe in marriage to girls as young as fourteen years old. Some of this is abhorrent to the rest of society. One of the draws of Mormon belief is that their literature asks for only a ten percent tithe or gift to the Church. This has become popular in some societies like the South Pacific Islands where, in the past, Christian missionaries insisted on much more to support the Christian way of life, making the missionary quite wealthy within that island's society. The statistics would indicate that there are approximately 14.8 million who follow Mormon religious practice, with about thirty percent practicing regular attendance at worship.

Also prevalent in Christian belief today are Jehovah's Witnesses. Their literature is the book, started in 1950 with the New Testament portion and fully published in 1961, called the New World Translation of the Holy Scriptures (NWT). They call this a translation of the Christian Bible, published by the Watch Tower Bible and Tract Society. It would probably be more correctly called an interpretation of the Christian Bible. It is used and distributed by Jehovah's Witnesses. Scripture from the Christian Bible is prevalent and is interpreted in a way that becomes meaningful to their needs. If you would like to look up some of the text in the Christian Bible, it will give the inherent belief structure of their system, and you may follow with your own investigation. They believe in God, but not the popular tri-person (Trinity)—God the Father, God the Son, and God the Holy Spirit. To them, God is separate from the son. In a brochure published by their foundation, Watch Tower, they postulate that the Council of Nicaea, during the reign of Constantine, Emperor of Rome in 320 AD, decided that the ambiguous term Trinity was a conciliation of his use of the term and that "Jesus was of one substance with the father," which laid the unsubstantiated foundation for the Trinity that became foundational in Christian doctrine. They use the following reference in the Christian Bible to support their published tenet of belief:

To us there is but one God, the Father (1 Cor 8:6 KJV)

Hear, O Israel: The Lord our God, the Lord is one. (Deut 6:4 NIV)

Jehovah's Witnesses say that they take Jesus at his word when he said, "The Father is greater than I am" (John 14:28 NIV). Therefore, they do not worship Jesus, as they do not believe that he is Almighty God.

Another practice among Jehovah's Witnesses is not accepting blood transfusions. Although they do not believe in faith healing, they vehemently abhor blood transfusions. For them, this is a religious issue, not a medical one. They say both the Old and New Testaments clearly command abstinence from blood (*c.f.*, Gen 9:4; Lev 17:10; Deut 12:23; Acts 15:28, 15:29)

Finally, we come to the primary literature of mainstream Christians, both Protestant and Catholic, that many others, including various sects of Christians, are contesting, (re)interpreting, comparing, and translating: The holy Christian Bible. There is no way for me to present this document in a way that learned scholars have spent several lifetimes studying. So, I give you the astute literary words of a scholar, and my good friend, who says it infinitely better than I ever could.

Pastor H. Paul Schmidt, MDiv

> *The Bible*
>
> The primary text for the Christian church is the Bible. The book, or more correctly, a collection of sixty-six books, is divided into two sections. The first section is called the Hebrew Bible or the Old Testament. The second part was written in Greek and called the New Testament. The various books of the Hebrew Bible were written over many years, probably centuries, often collecting material that was first a part of the oral tradition of the Jewish people. The New Testament, written within a century of the life of Jesus the Christ, is made up of eyewitness accounts of the life and teaching of Jesus. It also includes letters exchanged within the early Christian Church that seek to explain who this Jesus the Christ is and how his life, death, and resurrection affect human life and all creation. Many Christian bibles also include additional books that have various degrees of authority. These books are commonly referred to as The Apocrypha.
>
> H. Paul Schmidt, MDiv

When I presented my own description to another Lutheran educator and theologian, I was told the following: that my testimony was not Orthodox Christianity. I was told that God's gracious steadfast love is revealed in the Hebrew Bible, and it was all Jesus knew. Thank goodness that I am not trying to present myself as an interpreter or scholar about the Christian Bible. I would be subject to endless criticism and the opinion of others with numerous grandiose documents to prove their point. But my friend quoted here, H. Paul Schmidt, gives great literary insight into popular conceptions of the Christian Bible. Further investigation is available on the Internet with numerous interpretations by scholarly individuals.[18]

Sacred Texts and Types of Judaism

Judaism is the foundational theology of Christian belief. Jesus Christ was a Jewish rabbi, or priest. The origin of Judaism dates to Abraham, around 1986–1857 BCE, much like Islam, which is also attributed to the family of Abraham (one of his sons), but has its tenets based on Muhammad 2500 years later in the sixth century AD. Judaism, as we know it, really became foundational with Moses and the exit of the Jews from Egypt. This date is unconfirmed by historical research, so you decide— either 1440 BCE or 1290 BCE, as both are quoted by scholars. God communicated directly with Moses as he led his people toward the Promised Land—which turned out to be desert for a long time in the beginning of their journey. They finally settled in the Middle East and prospered. Through conflict, war, and occupation by other cultures, they eventually lost their homeland and scattered around the world. They eventually reacquired their lands after the Holocaust in Europe, where six million of them were exterminated. Depending upon what statistics you believe, there are between 14.5 million and 17 million people who practice the faith of Judaism throughout the world.

The Torah, as well as the Prophets (Nevi'im) and the Writings (Ketuvim), are all part of the larger book known as the Tanakh or the Hebrew Bible. Oral tradition is represented by later texts such as the Midrash and the Talmud. Remember that the history of Judaism was oral until much later in history, and rabbis and their followers could recite the Tanakh from memory. The Christian Old Testament is incorporated into the Torah, and the first five books of the Old Testament are the written Jewish Law of Moses.

Branches of Judaism include groups that have developed from ancient times. Today, it is often stated that there are three branches: Reform, Conservative, and Orthodox. However, several smaller movements also exist. These branches are mainly present in Europe and North America. In Israel, there is great consternation between non-religious and Orthodox Jews. I will never forget going to the Wailing

---

[18] "The Bible; So Misunderstood It's a Sin". https://www.newsweek.com

Wall in Jerusalem with a non-religious Jewish fellow as my guide and him explaining how to avoid the Orthodox (Hasidic?) Jews and forbid them from wrapping a ribbon around my arm. He was very vocal and adamant in his instructions. I have also attended a Jewish celebration in one of their synagogues, presided over by a female rabbi. This was an eye-opener because of my understanding that Jews separate men and women during worship. Clearly, there are modern versions of Judaism that do not feel compelled to obey ancient tradition.

Just like Christianity and Islam, Judaism also has fundamentalists. Most every male Jew wears a hat or kippah when they attend synagogue. However, some carry this to extremes, like the Hasidic movement, and wear curls and grow beards. Some fundamentalists even wear a "pill box" on their foreheads. Most Jews these days are non-religious and choose to only trace their culture back as a DNA spectrum.

The Sacred Texts and Types of Islam

Three important books exist for the interpretation of Islam.

> The two main sacred texts are: the Quran (also spelled "Koran") and the Hadith (or Hadeeth). These books teach and illustrate Islamic beliefs, values, and practices. They are also important historical documents (especially the Quran), which tell the story of the origins of the Islamic faith.

"Islamic sacred texts." Religion Facts.com. 21 Nov. 2016 Juliette Bently, Teacher, also http://www.religionfacts.com (Juliette Bently 2016)

The third book, the Sunnah, is also used in conjunction with the Quran. Interpretation of the Quran is literal, and the original text in Arabic is the only accepted document for the function and preservation of Islam

The Quran is "The Book" of Muslims, believers of Islam. Since Islam teaches about seven heavens, the Quran is considered to have existed in the highest (seventh) heaven but was brought down to the third heaven for some time. There the Angel Gabriel took it one piece at a time to reveal it to Muhammad, who is often called the "illiterate" prophet. During his lifetime, Muhammad recited it and shared his visitation with the people around him. Many in his company could quote large passages, some even the whole Quran, by heart. This is also prevalent in Judaism where learned rabbis and other Jewish scholars would and could quote from the Torah. There is and has been a heated theological debate about whether the heavenly copy of the Quran was created or has eternally existed and so was "uncreated".

The first assembly of the Quran was done by the first Caliphs—Abu Bakr and Umar. This necessitated transcribing verses that Muhammad spoke. When Muhammad died, the Quran had not been written and published. It was a collection of various verses written on various media like parchment, leaves, tablets, and sometimes even stone. The first Caliphs put these thoughts and snippets of verse together and published a basic Quran.

The second sacred text is called the "Hadith," which means a "narrative" or "report". It is a record of Islamic tradition: the words and deeds of Muhammad, his family, and his friends. It is second only to the Quran in importance for Muslims. Although not regarded as the spoken Word of God, like the Quran, the Hadith is an important source of doctrine, law, and practice, and is legally used in conjunction with the Quran as prophetic. Another way of viewing the Hadith is that it is the written sayings of the Prophet.

The third sacred text, the Sunnah, establishes custom, gives precedent, and dictates conduct and tradition, typically based on Muhammad's example. These actions and sayings of Muhammad are believed to complement the message of the Quran. Thus, it is the source for revealing Islamic policy, making it a primary source of Islamic law. In the legal field, a basic tenant of Islam is Sharia law. The Sunnah complements and stands alongside the Quran. Most Muslims believe that Islam is not only a religion but also a political and legalist way of life. The Sunnah is the written text of the way people lived and is believed to have been passed down from previous generations. Early Muslim scholars developed and amplified the concept of Sunnah to give a complete picture of Muhammad's life. It is based on the Hadith reports.

Islamic law is called Sharia. Sharia is contrived from the words of Muhammad used in the books of Hadith, Sunnah, and the Quran. Sharia Law is firmly fixed with expressions of how to live personally and publicly. Sharia Law is completely encasing with actions necessary for the legal system, public behavior, and private behavior. Even one's private beliefs are regulated by this law. Sharia Law is the most discriminating and demanding of any law in the world today. Women especially are confined to secondary roles, even regarded as property in the opinions of some in non-Islamic states. There is some latitude given by Imams in interpretation. Islam regards the world as divided into two separate states. One is the home of Islam, and the other is regarded as war. Those of other religions are only tolerated if under Sharia Law they provide payment to the Islamic leaders in penance, agreeing they are not the chosen people of God. Sharia Law is the product of the rule of Muhammad in 600 AD. However, the religion existed prior to this period and is considered the prodigy of Ishmael, the illegitimate son of Abraham and Hagar. Incidentally, Abraham is also the father of Isaac the founder of Judaism. No documentation appears to exist on Islamic Law prior to Muhammad. The only mention found regarding Ishmael is about the Haj and the *Kaaba* or black stone constructed as a place for pilgrimage, which was introduced by Ishmael.

Under Sharia law, it is acceptable for an Islamic man to marry a female who is not Islamic, providing he performs some task. However, it is not acceptable if an Islamic woman marries a non-Islamic man. This is punishable by death, under some interpretations. There are clearly exceptions to this within some of the more social movements of Islam, such as Ahmadiyya.

While I was in Tanzania for a month during Ramadan, I hired a driver to show us the Serengeti, which consists of several national parks that include most of the country. An Ahmadiyya Muslim, according to his religious practice, our driver was not able to partake in food or drink during daylight hours. This did not influence his conversation or befriending of me, though. During the very bumpy journey in Jeeps (a ride often called an African massage), he related his story of his daughter. She was very well educated and currently a judge in the Tanzanian Justice Department, presiding over a large area of Eastern Tanzania. She informed her dad, my driver, that she had fallen in love with a man who was Christian and wanted to marry him. She felt she needed her father's permission to do so. My driver explained to her that if she loved this man and he loved her, she could absolutely marry him with his permission. In some areas this could have cost her father dearly, and perhaps her and her new husband also.

My other story of Sharia law is about a family of refugees from Pakistan whom I supported. They were Christian and had to leave with no notice when the Taliban arrived at the front door of their home to kill them. The wife was active in politics while the husband was doing missionary work in the Christian faith. The husband

and wife, along with their two teenage daughters and four-year-old son arrived in my country and asked for asylum under the Geneva Convention. This was granted, but for reasons unknown, they were not offered the ability to work or any social assistance and were therefore destitute. I stepped up and became a valued intervenor in their life, so they could eat and have a roof over their heads. In addition, I befriended them for a while until they were self-supporting. Through conversation, I discovered that Pakistan is called the Islamic Republic of Pakistan, and they have Sharia law. Under their version of Sharia, any three Muslims can attest to legal authorities that a person has blasphemed Muhammad or Islam. This law of blasphemy allows someone to misuse the three-witness action to acquire whatever they wish from others, usually non-Muslims. This could even be that they desire one of the non-Muslims' children as a new wife, or their property or livestock. The penalty for blasphemy is usually severe, resulting in major consequences for the person accused, including the loss of life. Further, the imam of the region seems to have more power in this than the courts. Fortunately, the family I assisted survived and now live productive and normal lives in Canada.[19]

Hindu Sacred Texts

There are four divisions of the Hindu practice, just like in Christian practice whereby there are Roman Catholic and Protestant and thereafter many subdivisions. Each has its preferred literature and practices. They are called: Vaishnavism, Shaivism, Shaktism, and Smartism. Their names are derived from the gods—the main gods of the Hindu religion—whom they follow.

Goals of Hindu Life

There are four main goals in Hindu life. They are *kama, artha, dharma,* and *moksha*. These goals are understood as applicable to men only, and each is more important than the one preceding it. It is understood that it is okay to pursue them.

---

[19] For more insight from a person who is Islamic and has become a highly educated scholar, politician, and educator, try this interesting read: Ayaan Hirsi Ali, *Heretic: Why Islam Needs a Reformation Now* (New York: Harper Collins, 2015).

Kama

Kama is pleasure. It is about cravings of the mind and the physical body. It is human desire for passion and emotion. The Hindu god of love is named Kama. This name is the derivative for the famous Hindu guide to the erotic activity of love, known as the Kama Sutra.

Artha

Wealth and power: it's okay to want these. Pursuit of them is considered noble. A person needs them to raise a family and maintain a household.

This is especially true for the upper classes, or castes, of Hindu society. Artha, or wealth, is sought after to fulfill one's destiny. Under the caste system, some were made to be rulers and kings, while others were made to be beggars. Kings have every right to seek wealth and power; it's their duty and necessary to maintain a balanced society.

The opposite is true for the lower castes of Hindu society. It is permissible to seek to provide for their families; however, they should not seek wealth to move on up the social ladder.

Dharma

Dharma is a huge part of the Hindu faith. None of the other goals are as significant as dharma. Dharma means duty. It's a set of standards by which a Hindu should live. However, each person's dharma is different. It is required of them to accept their station in the caste system as part of their duty or dharma. It is also the duty that keeps people in their place within the caste system

Moksha

Moksha is nirvana, a state of perfect happiness that is the ideal or idyllic place. It is a state of eternal bliss and emptiness. It refers to the cycle of birth, death, and rebirth.

The Vedas

The four Vedas are the Rig Veda, Sama Veda, Yajur Veda, and Atharva Veda. Hindus use these as their primary texts. Since Buddhism, Jainism, and Sikhism are similar, the Vedas influenced those religions also. It is tradition that the Vedas were written at the same time as the creation of the universe. However, scholars have determined that the Rig Veda, the oldest of the four Vedas, was composed about 1500 BC, and put in order around 600 BC. It was finally committed to writing in approximately 300 BC.

The Vedas are ancient Indian hymns, incantations, and rituals. Along with the Book of the Dead, the Enuma Elish, the I Ching, and the Avesta, these religious books are among the most ancient texts on religion that still exist today. In addition to spiritual value, the Vedas give a perspective of everyday life in India four thousand years ago. They are also valuable in linguistics as they are the most ancient texts of their kind in the Indo-European language.

Types of Sikhism and Their Sacred Texts

I count eight sects within the Sikh religion: Udasi, Nirmala, Nanakpanthi, Khalsa, Sahajdhari, Namdhari Kuka, Nirankari, and Sarvaria. Clearly, they range from orthodox (fundamental) to varying degrees of liberal. Khalsa is the word I constantly see in my daily travels near where I live. I also know that Sikh people are in communities with others around the world, and they are accepted as highly educated, intelligent, benevolent, and altruistic. They generally desire to fit in with the community, and in so doing, support it generously. In my community, they are hard workers and entrepreneurs.

Sikh literature and teachings state that there is one creator, God, and that the Guru is central. A Guru is a teacher. Everything that teaches is then Guru, God: religious leaders, scripture, community. Sikhs share a belief in karma with Hinduism (pleasure), reincarnation, and an "ultimate illusion of the world". Dedicated (khalsa) Sikhs—we could also say fundamental—are distinguished by their uncut hair, comb, metal bangle, knee-length pants, and small kirpan (dagger). Sikhs oppose parts of the caste system. They also have communal kitchens in the basement of their temples open to all who desire to partake. There, people are encouraged to eat together. They are tolerant of other religious traditions, and Sikh temples symbolize this by having four doors facing each point of the compass, through which anyone can pass. It is expected that they abstain from alcohol and tobacco, and some do. Sikhism is not evangelistic, but people may convert. I have enjoyed my occasional visits to their temples near me where I have been openly welcomed, taught a little of their religion, and openly encouraged to visit again.

There is an apartment building near where I reside that has unfortunately placed a large permanent sign on its exterior stating, "We remember." It has a giant picture of Sikhs trying to disembark the *Komagata Maru* in Canada in 1914. This is an historical black eye, where North America was challenged by 376 potential Sikh immigrants over excessively restrictive immigration policies that were particularly negative to those Sikhs from the Punjab region. But history is selective by the proponents of the message.

So, I ask myself, "What has this Sikh religion got to do with this political event, allowing such a negative political statement to come from a few ordinary citizens of this Sikh community, and especially from people who were not even affected by this shameful Canadian decision?"

A personal negative reaction was brought on within me, an airline captain, remembering the much more recent unholy use of a bomb by this same community in June 1985 to blow up an Air India airliner over the Atlantic Ocean en route to India, killing all 329 aboard—and another that went off in Japan, killing one ramp worker. This was all in the name of Sikh retribution toward India for deeds committed in the Punjab region of India around that time. How does this horrible act of retribution toward India get justified in the minds of devout fundamentalist Sikhs? How does this get lost in the advertisement of "We Remember"?

Just by saying this I will now be branded as a prejudicial miscreant, even though I have a demonstrated history of aggressively helping those from the Punjab region obtain asylum in Canada, and my wonderful Sikh neighbors are the ones I have chosen to access my dwelling when I am away on world travels. This I ponder as I write about the literature of one of the world's major religions.

The Guru Granth Sahib is the book of Sikh scripture, or bible in another vernacular.

> *Compiled by Guru Gobind Singh, it contains compositions of six Gurus, namely Guru Nanak, Guru Angad, Guru Amar Das, Guru Ram Das, Guru Arjan, and Guru Teg Bahadur. It also has writings from people of other religious faiths. Hymns within it are also an important part of Sikh worship and are arranged by the thirty-one musical forms in which they were composed. The hymns that comprise the Granth were originally written in several different languages: Persian, mediaeval Prakrit,*

Hindi, Marathi, old Panjabi, Multani. In addition, there are Sanskrit and Arabic portions. This makes it extraordinarily difficult to translate.

The Granth is considered the living embodiment of the Gurus; it is called the "eleventh guru". It is treated with the greatest respect. This is the reason for the honorific titles that make up the full name of the book. There are protocols that are expected to be observed in while reading of the Granth. A Sikh reader would suggest the following: "Out of respect, it is advised that before you read, that you cover your hair." This is normally with a turban or a piece of cloth provided by the gurdwara.[20]From literature in a Sikh temple and attributed to;

The Sikh religion by Max Arthur
Macauliffe (1909)

This of course is the reason you see Sikh men wear a turban and women wear scarves over their heads.

Types of Buddhism and Sacred Texts

The types of Buddhism are Theravada, Mahayana, and Vajrayana, along with a fourth, Zen Buddhism, which grew out of Mahayana and has gained increasing popularity in the West.

The oldest form of Buddhism is Theravada, mostly from Southeast Asia (Thailand, Myanmar/Burma, Cambodia, and Laos). If you want to untie your English tongue, it could also be called "Doctrine of the Elders". I have been to Cambodia and Laos and have been the guest of a monk of this tradition. Most enjoyable of this time was the very early walk of the monks collecting their daily rations from the handouts of the villagers as they traveled in a parade-like line in the city of Luang Prabang. He also took us to his meager living quarters in the temple. Although Myanmar/Burma are attributed to this branch of Buddhist belief, recent events causing the displacement of the Muslim population therein are out of context from my point of view of the wonderful practice as witnessed in Cambodia and Laos.

---

[20]Max Arthur Macauliffe, also known as Michael Macauliffe, was a senior Sikh-British administrator, prolific scholar, and author. Macauliffe is renowned for his translation of Sikh scripture and history into English.

Mahayana Buddhism came from the Theravada tradition approximately 500 years after the Buddha attained enlightenment. Mahayana Buddhism focuses on the idea of compassion and espouses bodhisattvas—i.e., someone able to reach nirvana but who delays in doing so due to compassion for saving suffering beings. Several schools have evolved from Mahayana, including Zen Buddhism, Tibetan Buddhism, Pure Land Buddhism, and Tantric Buddhism.

Vajrayana was last in the evolutionary process and provides a faster path to enlightenment. The belief is that the physical influences the spiritual and that the spiritual, in turn, affects the physical. Within the practice of Vajrayana there are rituals, chanting, and tantra techniques, as well as a basic understanding of the Theravada and Mahayana schools.

Zen Buddhism originated in China, guided by the monk Bodhidharma. Zen Buddhism treats daily practice and zazen meditation as essential to attain enlightenment and minimizes rigorous study of scripture.

The Buddhist authority or rules (canon) consist of the sutras, which are the words and teachings of the Buddha. There are also several supplementary texts that teach rules of conduct and provide discussion on transitional states after death.

The first major book is the Tripitaka (Pali Canon), which has three divisions. Originally written in the Pali language, the first division is the Vinaya, which covers the rules of conduct for daily affairs for monks and ordained nuns. But it also gives the Buddha's reasons for the rule to maintain harmony within the large spiritual community. The second division is the Sutta, which contains all the discussions of the Buddha and his closest disciples about the central teachings of Theravada Buddhism. The third division is the Ahidhamma, where the Sutta discussions are reworked into a framework to investigate the nature of mind and matter.

The next major book is the Mahayana sutras. Mahayana Buddhists teach that enlightenment can be attained in a single lifetime, and this can be accomplished even by a layperson. Mahayana writings are Chinese Buddhist, Tibetan Buddhist, and Sanskrit Buddhist.

The Tibetan Book of the Dead is the third important book. It discusses *bardo* or the time between death and the next rebirth. The intention is to guide you through the experiences of death before rebirth.

Buddhism espouses reincarnation. The goal of a Buddhist is enlightenment (*nirvana*) and to be removed from endless reincarnation and suffering. Some see Buddhism as a religion, others see it is a philosophy, and others think it is a way of finding reality.

Unlike other Buddhists, Zen Buddhists don't dwell on sacred texts. The very nature of Zen Buddhism means going beyond intellect, logic, and language. It is an attempt to understand the meaning of life through meditation. This is a parallel to the mysticism found in original Christian practice, wherein Christians prayerfully contemplated scripture.

Atheist Writings

There are no comparable texts to choose from. Atheists have nothing they are asked to "believe", and no atheist author asks them to believe anything that he or she may write.

Religious texts such as the New Testament and the Quran both contain rules for ritual as well as moral mandates. Atheists do not have any moral mandates, nor do they have specific atheist rituals. The point of atheism is it uses an appeal to reason, rather than an appeal to authority, so it does not have an authoritative book. An atheist would note that calling atheism a religion is flat out false. It's a belief they have in common, not a list of rules to live by.

However, when pressed, an atheist would refer anyone to the chronological time line of their belief starting with the ancient Greek philosophers like Democritus, Leucippus, Protagoras, Epicurus, and Diagoras (c. 448–388 BCE), a.k.a "Diagoras the Atheist of Melos", who was a disciple of Democritus. All these philosophers were deep thinkers in physics and astrophysics, far ahead of their times. Sir Isaac Newton is purported to have been a student of the school of Democritus and took those early ideas into deep present-day thinking. I will provide more on that in the sections on science and philosophy.

Stop, Think, Discuss

This explains the basic theology of most popular religions, practices, or beliefs. Our questions tear down the statements for discussion. This is good practice, but don't forget how to put everything back together again with appropriate reasoning when determined!

We should also debate what the hand of man creates as an act of theological perspective, purportedly from direct communication from or superior knowledge of God or the gods. Have the various rules and practices within each religion been generated by man or by God? Also, note the date or timeline when/if this occurred and why, to help put everything into perspective with the political situation of the times.

Question one: Do you see anything so far in our descriptions that could be interpreted as made by the hand of man?

Question two: The history of almost all these theologies dates back several thousand years. Do you think they were a product of man's ingenuity or God-created? If God, then why would God create these and have man transcribe them?

Question three: What appears to come first in this? Was it an idea created by God or by man?

Question four: Was there a precursor that caused the creation of each of these religions? If so, what was it? What is the wonder and mystery in all these religions? Is mystery necessary?

Question five: How would you describe these religions in your own words?

Question six: Are there issues in these descriptions of each religion with which you disagree? Has Harry Potter— creative thinking—taken over the dialog on this issue?

CHAPTER 5
The Hand of Man Creating and Recreating

In the history of *Homo sapiens* on this planet, humans have created, destroyed, and recreated everything they can control or get their hands upon. With rare exceptions, like Mother Theresa, this has been an exercise of gaining personal control over the process at hand. Most often, this control is for personal use, ego, financial gain, elevated perception of self-worth, control, ability to direct or influence others, exaggerated opinion of self-worth, or ability to influence others. This is true for every aspect of human endeavor, including theology and science.

Since this is a chapter on theology, let's try to review how the hand of man created, changed, morphed, and recreated theological ideas and teachings to support the opinion of the person(s) changing the practice. The whole exercise is to ask where the God stuff exists and where the human stuff takes over and why. I will restrict this to the two most populous theologies on the planet: Christianity and Islam.

The Evolution of Christianity

As previously discussed, Christianity is a progression of the Jewish faith that includes Gentiles. During the hundred years following the crucifixion of Christ, the Jews were severely abused, punished, and marginalized as they rebelled against Roman rule. They fought the Roman Empire over taxation, and Rome mistreated the citizens of Israel. The Jews were so vexed that they were quietly looking for leaders to lead a rebellion; many even felt that Jesus Christ was the one who could take them out of this state of occupation and mistrust. Jesus' God-given authority was to be different, and he presented a way of correcting this abuse of power through love, not aggression. It is a study in human dynamics that has lasted to this very day, unprecedented and unmatched in its simplicity and power.

Apostles and followers of the Rabbi Jesus of Nazareth became the second most popular movement in the Jewish tradition in Jerusalem. They advocated an evangelistic, inclusive approach to their version of the Jewish faith. At about 70 AD, Jerusalem was overrun by 70,000 Roman soldiers, and the Jewish temple was destroyed. Titus, the son of the Roman emperor, was sent to ensure compliance of the Jewish rabble revolting and causing so much trouble. Most of the Jesus sects of Jews remaining alive were Gentiles. The Gentiles, not having been included in the Jewish traditions since birth, had a problem with the Jewish religious rite of circumcision. They loved the mystery and miracles of Jesus and his followers and the treatment of women and the poor.

The other religions, including paganism and other Jewish sects, did not try to encourage anyone else to participate in their belief. And since the Jesus movement was also persecuted, they gained a measure of popularity. Also, since this movement supported the value of women and used miracles to gain popularity, it acquired quick recognition.

All but one of the Apostles were put to death by the Roman authorities. The book of Acts in the Christian bible gives a good history of the early Jesus of Nazareth movement, which had previously been called The Way and eventually became Christianity.

Nearly three hundred years after the crucifixion of Jesus, Emperor Constantine was going into battle against another Roman general, and the odds were not favorable. The evening prior to battle, he saw a vision of a burning cross in the night sky and heard the words, "Through this sign you shall conquer." Constantine won the battle and became a convert of the Christian movement.

In 315 BCE, Rome was slowly decaying, and this is one of the ways Emperor Constantine used to keep Rome in power. Constantine conferred a gathering of Bishops in Nicene in 325 BCE, where they discussed the nature of Christ, and from that gathering created the doctrine of this new faith that included the creation of the Trinity—the Father, the Son, and the Holy Ghost. They did not create Christianity but transformed the philosophy of Jesus who preached poverty, love, forgiveness, grace, and nonviolence into a political ideology of power that promoted unification with a cohesive strategy that was necessary for the reunification of the conflicted Roman Empire at the time. Faith begat blind allegiance, which precluded reason through unquestioned belief and which also, through this new structure, justified the station of the Emperor. From these Roman roots, the Roman Christian Church was born as an institutionalized faith mostly removed from the control of Jewish tradition. Several other gatherings of the leadership of the Church were held over the years, and a group of Orthodox Christians was not included in some of these dictates. They formed the Orthodox Christian Church, which moved to the north and west of Rome.

Christianity like all religions became a tool with rules made by man, which took something wonderfully valued and offered it as spiritual, including benefits through cooperation, within its mandate providing you chose to follow their set of truths or values. Then, progressively, the hand of man misconstrued the values to gain power, wealth, and fame.

Evolution of the Roman Catholic Church

The leaders of the Catholic Church became people-centered over the next one thousand years. They also gleaned power for themselves, which was likely also Constantine's purpose in his original absorption of this movement in approximately 315 AD. The pope became all-powerful, and the bishops became so powerful that they had the power to put to death those who were against their rulings, destroying people and teachings that did not adhere to their own mantras. To avoid inheritance issues, the emperor made the clergy practice celibacy. That way, they could have most of the power of the emperor without Rome losing its wealth.

The Earth was the center of the universe. Any ideas that were different from those of the authority were immediately dismissed and even those behind the ideas were called heretic and often burned at the stake. The Church became pageantry, pious, secular, centrist, unthinking, unfeeling, hierarchical, and manipulative. Schools of learning from the past were destroyed and all their books burned. As John Dalberg-Acton[21] said, "Absolute power corrupts absolutely."

A wonderful fictional book was written on this subject by Ken Follet. *The Pillars of the Earth*[22] is the first in a series of several books wherein he goes to great length to replicate the atmosphere of the Roman Church in England back then. To me, this book is a modern classic and should be mandatory reading for everyone. It gives visual perception to the arduous life of living under the rule of the Catholic Church prior to the Reformation.

[21] John Dalberg-Acton (1834–1902) was an English Catholic historian, politician, and writer.

[22] *The Pillars of the Earth*, is a historical novel by Welsh author Ken Follett published in 1989 about the building of a cathedral in the fictional town of Kingsbridge, England.

Finally, about 500 years ago, the Industrial Revolution came into play. The people of the Western world, Europe, were dissatisfied with the power and use thereof by the Church. Within the Church, two powerful people were instrumental in breaking away from the norms of the Catholic Church. Martin Luther, a German priest, declared his distaste for all the rigors of the Church in 1517, and shortly thereafter, John Calvin did the same in Geneva and France in 1535. Both reformers had the same objective but went about it in different ways. There is minor conflict of interpretation between the two, even to this day. Both became Protestants, a division of the Christian faith. Both have the lay people able to talk directly to God, allow the Bible to be printed in their own language, and did away with much of the legal restrictions and pageantry of the Catholic Church.

From that time to this, the Protestant Church has evolved into myriad branches, just like Islam. There are branches that want to go back to old practices, and others who are hardly even recognized as Christian. Some want to recognize the Bible as literal, others as metaphor. Some want a liturgical service (with a strict order of service); others want more freewheeling, with a leader giving an opinion. All of them recognize life after death, sin, and evil. All of them have the Bible as their primary source of teachings and history. Protestants recognize the failure of the absolute power of the Church and separate Church and State in their politics. There are some, though, that would have the Christian faith become a vital component of state law, with the power to influence government.

In recent years, theological agreement has been achieved between Reformed, Catholic, Lutheran, Presbyterian, and Anglicans through the doctrine of "grace through faith". As with and following the discussion on Islam, this points the discussion toward liberal, fundamental, and extreme religious thinking.

How Has Islam Changed Over Time?

Islam has gone through many changes over the years, and there have been many movements within the religion that have sought to reform it.

First Major Reformation—Quran By Usman

As previously mentioned, the writing of the Quran is attributed to Muhammad, who was given the verses by the Angel Gabriel. After Muhammad's death, the Quran wasn't in the book form that we see today. This was not satisfactory to Caliph Usman (aka Uthman, the third caliph after Muhammad), who ordered a proper composition (from 644 to 656 AD) of the Quran into what is seen today. He ordered all older versions to be destroyed to ensure that all existing versions were the same. Islam thus experienced its first transformation.

It is stated that the Quran, as seen today, is exactly like the version that Usman commissioned. This version is the one referred to in legal authority to resolve disputes of interpretation.

Second Major Reformation—Hadith

The directives of Hadith are expected to be followed by Muslims around the world. Any changes to the Hadith mean a change in the way Islam is practiced. The interpretation and authenticity of Hadith is one of several issues on which Sunni and Shia Islam differ. The Hadith is also one of the practices of Islam that provides room for interpretation. These differences are the issues of opinion that lead to creation of many different schools of thought in Islam.

We've already spoken of the schools of thought in Islam, which are: Sufism, Quranism, Ibadism, and Ahmadiyya. All are social in context. Then there are all the political movements, like Salafism and Islamism, which declare Islam as not only a religion but also a political system. Further, there is the idea that their laws should politically govern all the legal, economic, and culture matters of the state.

Rulers Reform Islam in History

Islam has been the state religion for several dynasties. Depending on the ruler, it also underwent major transformations. Of course, during these various dynasties, it has depended upon how closely the intent was implemented and how literally the verses of the Quran were interpreted. During the period of major scientific progress from the eighth to the fourteenth century, Islam was progressive, and was also so during the period of extensive political reformation during Akbar's rule in India (1556–1605). Islam has also been intolerant and hostile toward other religions, and even toward other branches of Islam. One example is during the reign of Aurangzeb, a descendant of Akbar; another is during the Taliban rule in Afghanistan or even in the modern Kingdom of Saudi Arabia and another now in Africa and in several countries of Islamic origin. In this process of modification and change, Islam is the same as every other theology, especially the example of Christianity.

Once again, this brings us to a discussion on liberal, fundamental, and extreme types of thinking in theology.

What Is Liberal, Fundamental, and Extreme in Religious Thinking?

Liberal religion is that which embraces the evolution of religious thinking. It does not rigidly follow a single authority or writing. Because it understands how ideas and knowledge change with time and research, it is not rigid, and can draw from many sources, even other religions. It recognizes that original teachings could be metaphors for better understanding. It accepts today's understanding of knowledge and uses that knowledge to change religious understanding and modify fundamental belief. Liberalism does, however, only accept that there is a God. It is ecumenical[23] if Christian, and cooperatively agreeable if not Christian.

Fundamentalism in religion is characterized by strict adherence to a standard set of principles and texts. It has a strong distinction of difference of who is in and who is out regarding their own dogma. This gives the security of knowledge, within the group, that they are the only people saved by God. Fear-based, they often require acts of piety that signify their adherence to this devoutness of a true religion. Often, they desire to return to a previous iteration of their faith. Rejection of diversity of opinion is mandatory and universally held within the group. They also often require physical designation or segregation from the rest of the populace by the wearing of garments or icons that signify their piety. Fundamentalism is a part of every religious practice—Christianity, Islam, Hinduism, Judaism, and even Buddhism. Numerous sects of every theological practice have migrated to North America for the religious freedom tolerated there.

Extremism in religion is also called fanaticism. It is adversarial in nature. It is the unwavering belief that you are either for it or against it. Radicalization is the process of making an extremist point of view. The best examples of this are the cults that exist in modern Western countries. The examples best known are the Charles Manson Cult in the United States and the ISIS movement in Islam. Many others

---

[23] "Ecumenical" means promoting or relating to unity among the world's Christian Churches.

also exist, often masking their nefarious intentions as religious belief. Currently, in the Islamic community, but also in other religious communities, extremism is the desire to gain all power over the land and people within it for their own purposes. Death is often used to weaponize their belief system, even to others within their own faith community.

The Historic Overlapping of Christian and Islamic Territories

Then there is the issue of the Christian Crusades. Major issues cited concerning the Crusades involve urban legends surrounding them. It is often incorrectly stated by people who come from a background of generations of Christian thought that Muslims were the innocent party, the victims; that the Crusades instigated Muslim hatred of the West and, further, that Crusaders slaughtered innocent Jews and even other Christians. Also, there is the belief that crusading children were sent to war and that the Crusades were just another aggressive way to get rich.

Perhaps the worst misconception of all is that the Crusaders were held unaccountable because the pope promised them forgiveness of any sin committed while on Crusade. Like most urban legends, this fake news is based on only wisps of information, usually misunderstood and/or misrepresented. It has become the purview of the "politically correct" citizenry who reside in their mostly Christian environment, where criticism is more openly acceptable.

So, let's review the nine Crusades, from 1096 to the last one in 1272.

Islamic belief really came into its own with the introduction of Muhammad to the world in 570. With his death in 632, he left behind a profoundly changed Islamic doctrine. Doctrine is used here as meaning to define a stated policy of affairs upon which you might act. For instance, Lutheran doctrine is the policy through which Lutheran theology is enacted. I will summarize some of the Islamic doctrine that would apply to this discussion. At the time of his death, Muhammad was waging war with Mecca, which, back then, was mostly wealthy and Christian.

The mindset of traditional Islamic belief in those days was there were two regions in the world. One was the state of Islam and the other was war. This concept is not totally original in the history of our world. At the time of Jesus Christ, the "chosen (Jewish) people" saw the world as one of two places: Jewish or everywhere else.

The spread of Islam was by war or war-like activity. A devout Muslim would pray five times a day, indicating his steadfast conviction of his faith. This psychologically reinforced his actions on an ongoing basis, and even hour by hour. The theology of Islam is that you are held accountable for your actions and everything you do in the name of Islam is added up at your demise. This is almost like a mathematical calculation of positive Islamic deeds at the end of your life, and only they count toward salvation. Even better, you will be rewarded for excellent performance. The result of all this is that there is little compassion, grace, or caring in the traditional Islamic heart except if you are a guest in their house or land.

The Christian Eastern Roman Empire was in decay and finally fell to the Ottoman Turks in 1453. But it had constantly been under siege by Islam from the time of Muhammad. The armies of Islam fought quickly and mercilessly conquered Egypt, Syria, and Palestine. The Christian Western Roman Empire had fallen to the Germanic people in 476 AD, so there was a split in the leadership in Rome and little support for Eastern Orthodox Christianity, even though the Viking/Germanic heritage of the Western world accepted and promoted Christianity in their own way during and after conquering the Emperor of Rome in 476 AD.

The taking of holy lands and Jerusalem held by Jewish and Christians in the East was an offensive/defensive act between the aggressor Islamic armies and defensive (Christian) Roman and Jewish armies. The Islamic armies were winning, and after quick initial success, were slowly taking over the entire area occupied by the Roman Christian empire. For instance, by the end of the eighth century, they had taken over all North Africa and Spain. There was no assimilation. It was brutal and final. Stories of the Holy Land and Jerusalem being closed to pilgrims and of the brutality of the war were making their way back to the Western Christians who were now in control of the West. Although they were concerned, they did nothing in support of the now-occupied Christian and Jewish populations until 1096 AD when their pope, Urban II, asked them to open a way to the holy lands for religious pilgrimage.

Then, from 1096 to 1102 (the First Crusade), Pope Urban II asked Christians of the West to assist the defenders and open a route to the holy lands and Jerusalem. In 1099, Christian armies took back Jerusalem and brutally disposed of all their enemies. This is often called the most successful Crusade by European standards. In all, there were nine Crusades. The Muslims won the second one nearly 50 years later, and then won the third. They made heroes out of the Muslim leader Saladin and the English king, who became known as Richard the Lionheart. They made a truce to cooperate in Jerusalem, and Richard returned to England. The Ninth Crusade ended in 1272 after one year of war. Henry III was head of state in England, and his son Edward led the Crusade.

Besides being asked to defend their faith, why did they do it? In the wake of the Enlightenment (1685–1815), it was often stated that Crusaders were idle, worthless people who took advantage of an opportunity, robbing and pillaging in a faraway country. Discarded were the Crusaders' declarations of piety, self-sacrifice, and love for God. It was believed that these statements were obviously a sham and a disguise for darker designs.

Modern scholars have discovered that crusading knights were mostly wealthy, with their own land in Europe, which they willingly left behind to partake of this holy mission. Crusading was expensive and even the wealthy potentates of the time could spend their entire financial empire by joining a Crusade. Their only reason was to appear pious in the eyes of our God. There was no lust for material wealth. They were keenly aware of their sinfulness and eager to undertake the hardships of the Crusade as an act of charity and love. Europe is littered with thousands of medieval charters attesting to these sentiments.

Stop, Think, Discuss

Some people state that they want nothing to do with religion because it is the cause of war. Certainly, in the case of the Crusades, it is. This was also the case with the Hundred Years War in Europe.

Consider the wars with the greatest loss of life. Do you think religion was the cause?

I refer to the First World War, the Second World War, the American Confederate War, and others with less loss of life but wars nevertheless: the Korean War, the first and second Gulf Wars, the Spanish Civil War, and so on. Were these wars caused by religion or by something else, like lust for power, lust for riches or resources, and/ or ideology?

Question one: What is your opinion of the definition of liberal moderate and extremists in religion, as stated in this book? Give examples.

Question two: Considering the history of Christianity and Islam, as read here, what is the truth in your mind and want is perceived but not proven? What is the *ontology* of these statements? That is, what is the actual information, free from our personal bias, of these statements? And what is the *epistemology*? That is, what is the knowledge, as learned by us, including our preconceived notions, of these statements?

Question three: What appears to be real in this? Was the evolution of religion created by God or by man?

Question four: Absolutely everything that happens has a precursor that makes it happen. Then the event happens. What do you think caused the events that changed the religions of the world? Also, could the changes have been prevented or should they have been prevented?

Question five: How would you describe religious evolution in your own words?

Question six: Are there issues in these paragraphs with which you disagree? If so, what? Discuss. Consider the recent murders in the Jewish Synagogue in Philadelphia. Muslim mourners came as support and were welcomed inside. Explain your thoughts on this act.

PART TWO

THE HISTORY OF MANKIND, CULTURE, AND POLITICS

To learn, one must not only read but must also think.

Thinking is best done through discussion.
Chris Pedersen

Stop anywhere it would be prudent to ask any question.

PART TWO

CHAPTER 6

The History of Mankind, Culture, and Politics

Life

It was about 13.5 billion years ago that plasma, time, and energy gathered into one big convulsive event called the big bang. Then, 380 million years after that, matter gathered into constellations, individual stars, planets, asteroids, and all other matter. This is where the universe started, and without the universe, there is no man.

What happened one-millionth of a second before the big bang? Nobody can tell you and likely never will. What can be told with authenticity is what happened after the big bang because of our science and the research our scientific scholars have accomplished.

This is about the history of humans upon this solidified plasma called Earth. Astrophysics is in another chapter of this book.

About 3.8 billion years ago, atoms started to form into microorganisms, which are the precursors for the science of biology. I have not seen a stated scientific explanation for atoms that would form inanimate objects, like granite, to somehow randomly and suddenly form a complex microorganism that was life. Some would call this coincidence, others would call this a miracle, and yet others would call this an act of God or of a God who works through an evolutionary process, which is love—love of this planet, us, and all things on it.

The microorganisms modified over time into verdant life and animal life. Dinosaurs roamed the planet for millions of years. Finally, in the process of evolution, upright beings evolved; in fact, four types of two-legged upright walkers were present two hundred thousand years ago. Scientifically, we know of the species *Homo sapiens* because of the discovery of a female skeleton found in Ethiopia, Africa, dated back to that time. The three other upright beings present on Earth were: *Homo rudolfensis* (East Africa), *Homo erectus* (East Asia), and *Homo neanderthalensis* (Europe and Western Asia). But I am moving forward too fast.

Microorganisms need oxygen. They need water. There was no oxygen on Earth at creation. The water we encounter today, it seems, must have been delivered long after Earth formed. Or maybe not? It depends upon which scientific theory you read. Many scientists say asteroids and comets are the catalysts of this miracle. Recent studies say that Earth's water most likely accreted at the same time as the rock. So, you get to be the scientist on this one.

There was no oxygen. All agree. Within the water, cyanobacteria (or blue-green algae) were the first microbes producing oxygen by the process of photosynthesis.

What occurred that enabled cyanobacteria to enter our water and take over the production of oxygen? These microbes were life and came from somewhere or were created by some as-yet-unknown force.

There is no test to provide the precise oxygen content of the atmosphere at any point in Earth's history from the geologic record. However, one thing is very clear: The origins of oxygen in Earth's atmosphere derive from one thing—life in the form of microbes.

The First Miracle Upon Earth: Water

Water is the universal element for biological forms on planet Earth. Let's pause a moment to look at the miracle of water. Water is the only element that exists in nature in all three states of matter at the normal temperatures on Earth (solid, liquid, gas). It also covers 70 percent of the Earth and composes roughly 78 percent of the human body. It is the universal solvent. Leave any biological element in water and it will dissolve, in time.

Water in our bodies is another small miracle. Let's review this in more detail. Within every living thing, including plants, molecules and organisms need to move throughout its system, and it uses solutions for this. The best example is the oxygen in our blood stream. Some molecules need to move through membranes and use a process of diffusion and osmosis. These two processes are just another example of the vast, almost magical properties of water within our environment.

Some chemical reactions in the body require solutions to introduce reactants. Since water dissolves these reactants, it becomes a solution, important to the functioning of the body. Examples are the breakdown of carbohydrates and proteins in the digestive process. This water, which includes solutions, acts as a lubricant.

Capillary action, yet another characteristic of water, is the ability of water to flow in confined spaces without the assistance of, and in opposition to, eternal forces like gravity. An example of this is the drainage of the constantly produced tear fluid of the eye.

The two elements that make water—hydrogen and oxygen—have important and vital values that support life.

The hydrogen bonds give molecules of water cohesion and surface tension. You can see cohesion if you slowly overfill a glass with water and observe the water molecules holding together as a bulge of water above the rim. Only when gravity overpowers the bond will water spill down the side of the glass. This property of cohesion is what gives trees the ability to bring water up for the nourishment of the leaves, which are hard at work absorbing carbon dioxide and producing oxygen using the chemical reaction called photosynthesis.

Surface tension on water is observed when water spiders literally walk on the surface of water. The water molecules on the outside align and are held together to create an effect like a safety net made exclusively of the water atoms of oxygen and hydrogen.

Water is a component part in numerous other reactions, either as a reactant or as a product of the reaction. For example, look at digestion and aerobic respiration. When water reacts with a chemical to break the chemical into smaller molecules, the reaction is called *hydrolysis*. When water is formed as one of the byproducts when two molecules join, the reaction is called *condensation*.

Every living thing requires water to survive. Without water and the properties just discussed biological life would not exist. But that is not the end of the story!

If you are not impressed with these magical properties of water yet, hang on. Here comes what is probably the best part. Shortwave radiation from the sun is absorbed by and causes the water to change to a gas. It captures the sun's heat energy, which is retained within the molecules of the gas for a later time in the cycle.

This gas rises in the atmosphere, creating an area of low pressure. As the water vapor rises, it cools at a rate of three degrees per thousand feet until it reaches a temperature whereby it becomes one hundred percent saturated and condenses again, making cloud. By condensing, it gives off heat, and now the rising air in the cloud cools at a rate of saturated humidity, which is only one and a half degrees per thousand feet. This process causes water droplets to form around particles of dust, and we see it as rain.

Rain clears the air of dust particles, since it needs a nucleus to form on as it condenses into liquid. The seawater that originally escaped into the air as water vapor is now fresh water, and we can drink it, water our crops, mix it with other elements to create other important products, and sanitize our hands with it.

The low pressure that was created causes other heavier air to move into the area to replace the small depression created by the process. Areas of high pressure move into areas of low pressure, and thus we have wind. Wind moves in a circuitous route because of the effect of the Earth's rotation called the Coriolis force. This, in turn, causes moist, humid air to be forced up the side of mountains, compelling it to condense and create even more rain.

Large areas of air become either warmer or colder and thus lighter or heavier (cold being heavier). Heavy air moves along the surface into warm, light air, and the border between the two becomes a weather front with all manner of rain, wind, lightning, and so on. Even the lightning nourishes the soil with positive nitrogen as it causes electrons to move between the clouds and Earth. Without this cycle of life, started by water vaporizing, our planet would be akin to Mars.

Water also has the property to be a heat sink. It absorbs the heat of the sun in the ocean to a depth of about one hundred feet due to waves and currents. It slowly gives this heat back during the evening or when atmospheric conditions make the surrounding air cooler than the water.

Although water can expand due to heat, it is by a very small amount. The coefficient of expansion of water is a 0.000214 percent increase for every degree Celsius. This means that 100 feet deep of ocean water with a temperature increase of 15 C will increase in volume by 0.321 percent or 3.85 inches.

When water freezes, the molecules slow down in activity and increase the spaces around them, which increases its volume. Therefore, because it has more mass than weight as a liquid, it floats. It is the only matter on Earth that can exist in three states—water, liquid, and gas—at the normal temperatures found on the surface. As a solid, it retains its position until melted and provides storage of water for future use.

It also has the property of sublimation. Sublimation is the ability to transform from one state to another, skipping the intermediary step—solid to a gas, for instance. Hoar frost is an example of sublimation.

When you examine the properties of water, you see the engine that drives animal life, vegetation, and makes life possible on this planet. We see it every day and take it for granted, but it is one of those miracles that we humans constantly look for but never see.

Stop, Think, Discuss

Water occupies very close to 70% of the surface of the Earth. Yet only 10% of that water is fresh, potable water—the potable water that is so important to life on this planet.

There are excellent environmental movements that are seriously concerned with stewardship of the planet but, unfortunately, seemingly disregard preservation and cleanliness of potable water, in large part.

Can you think of some way to identify the need for good stewardship of potable water?

Question one: Is God a part of this discussion? Can you see any coincidence of evolution that may have caused the components of the discussion?

Question two: What is the *ontology* of these statements? That is, what is the actual information, free from our personal bias, of these statements? And what is the *epistemology*? That is, what is the knowledge, as learned by us, including our preconceived notions, of these statements?

Question three: What appears to come first in this? Was the formation of oxygen and water a creation by God or by coincidence?

Question four: Absolutely everything that happens has a precursor that makes it happen. Then the event happens. Do you have any suggestions as to what might have made all this happen? What, in your opinion, is the cause of life as science now understands it in the creation process?

Question five: How would you describe this process of creation, in your own words?

Question six: Are there issues in these statements with which you disagree? Has creative thinking taken over the dialog on this issue?

CHAPTER 7

Our *Homo Sapiens* Ancestors

The Nomads

A good definition of *Homo sapiens* has been suggested as "one who eats cooked food". Cooked food prevents sickness and disease, by processing out the viruses and bacteria in the food by means of fire. Not very scientific, but real. No other species of animal does this, obviously.

Early in CHAPTER 1, we talked about the skeleton of a *Homo sapiens* female. She was found in Ethiopia, Africa. DNA analysis of this skeleton was most revealing.

The mitochondria component of DNA within the placenta surrounding a baby in a mother's womb is always created only by the mother. No male influence is present in this chemical miracle, supporting and nurturing the baby. This component is passed from mother to mother, forever.

The mitochondria found in the skeleton of this female is identical to that found in the placenta of modern-day females giving birth to their families. Mitochondria is only passed from female to female in the process of the creation of an infant. This has been proven beyond a shadow of a doubt.

That means that our first mother, identified at this stage of our history, is this female found in Africa, and she lived two hundred thousand years ago. She has become nicknamed "Mitochondrial Eve".[24] This is significant in the history of our species. But also, the question becomes, "What happened to the other upright-walking humanoids present at the same time?" Approximately one hundred thousand years ago, Sapiens moved to the Middle East, where Neanderthals were predominant.

[24] Science Daily, https://www.sciencedaily.com The most robust statistical examination to date of our species' genetic links to "Mitochondrial Eve"—the maternal ancestor of all living humans—confirms that she lived about 200,000 years ago.

Sapiens did not last, were driven out of the area, and were not heard from again for about thirty thousand years.

But things change. About seventy thousand years ago, the XY chromosome of the male made a distinct change in the Y chromosome. Every male gets an X and a Y chromosome, whereas females only get two X chromosomes. This change was observed in the *Homo sapiens* species with the skeletal remains found of a male who became known as "Y-chromosomal Adam" within the scientific community.[25]Overnight, there was something significant happening to the *Homo sapiens* people. Once again, they migrated into the Middle East with the Neanderthals. But this time, they stayed. They were not driven out. Scientifically, this has come to be called the "Age of Cognition", seventy thousand years ago. Our ancestors suddenly became interested in exploring, invention, and community. They continuously moved from place to place, eventually taking over regions previously occupied by the other upright beings. They were excellent hunters and gatherers. No one knows how or why the other species vanished. By war or by assimilation? Regardless, they all disappeared.

Sapiens developed gods, religion, legends, and myths, which were a way of binding them together beyond the family unit without the conflict that would naturally arise outside the family unit in large groups. The language of gossip is the key factor in binding the whole group into one. Scientists know from research that gossip in language is a great influence on binding groups together. They also know that the greatest number of people each of us can effectively know, and be able to gossip about, is about 150 persons. Below this number, there is no need for a formal structure of ranks; legal and hierarchy in structure are irrelevant. So Sapiens were able to develop religion, government, and social structure with a binding legal oversight. Likely superior, this language skill is the tool that brought the end to the other upright beings, but since they no longer exist, we will never really know beyond speculation.

The Neanderthals also occupied what we now call Europe. Neanderthals were larger in the frame and more muscular than *Homo sapiens*. They also had a significantly larger brain cavity. It is known that as *Homo sapiens* immigrated to Europe, they lived together with the Neanderthals. Even intermarried. Today, if you have

[25] Science 31 Oct 1997: Vol. 278, Issue 5339, pp. 804-805 https://science.sciencemag.org Scientists have been searching ever since for "Adam", the man whose Y chromosome was passed on to every living man and boy. Now two international teams have found the genetic trail leading to Adam—and it points to the same time and place where Mitochondrial Eve lived.

your DNA tested, you will find that you have a Neanderthal element within your DNA string. So, the conclusion is that the Neanderthals did not disappear due to war with *Homo sapiens*.

I have had my DNA identified and much to my surprise I have an extremely large component of Neanderthal DNA in my system. So now I promote the fact that Neanderthals had larger brain cavities, larger frames, and were generally stronger than *Homo sapiens*.

So, what is it that made *Homo sapiens*, us, able to outlive every other upright-walking species? Historians on this subject would tell you that it is because *Homo sapiens* were able to construct a way of living together in a large community, peaceably, and that *Homo sapiens* created "myths" of justice, government, religion, social acceptance, and so on. All these qualities were the binding glue of large numbers of people living together, whereas the other species, like Neanderthals, were able to live together but not in larger groups. Thus, the Neanderthals were assimilated by the *Homo sapiens*.

Homo sapiens also became able to create superior tools for hunting, and they were excellent hunters. This is demonstrated by the fact that as *Homo sapiens* moved into other lands and continents, they wiped out all the large animals within those regions within two thousand years of their arrival. It is estimated that about half of the large beasts of the whole world were extinct by the time of the Agricultural Revolution. By example, in North America, the giant four-foot-tall beaver, the large woolly mammoths, and the saber-toothed tigers were all eliminated. South America lost forty out of fifty large marsupials to extinction. Ditto in Australia, where *Homo sapiens* wiped out the giant marsupials that existed.

Trade developed among the Sapiens. Evidence demonstrates that people using tools created from deposits not available in their occupied lands were nevertheless using these minerals in their manufacture of tools. Some of them had to have established navigation and trade routes. This is not evident in any other species on this planet.

It is understood by some scientists that Sapiens were the best hunters, and they did not have to work at finding food for more than about two to three days per week, and even then, it was easy with their capacity to learn languages, technology, and skills. Just like us today, they overdid it, and then they were in a food crisis from over-hunting and a population explosion. Sapiens are a social animal and that causes a baby boom.[26]Suddenly, the Change

Sapiens also can and do revise behavior rapidly. This is another trait that came with the cognitive revolution seventy thousand years ago. An example, given by Yuval Noah Harrari, and quoted below, gives this insight. He is trying to establish that Sapiens also change through behavior, not always from mutation. I insert it for controversy.

> The Catholic alpha male abstains completely from sexual intercourse or raising a family. This abstinence does not result from unique environmental conditions such as severe lack of food or want of potential mates. Nor is it the result of some quirky genetic mutation. The Catholic Church has survived for centuries not by passing on a "celibacy gene" from one pope to the next, but by passing on the stories of the New Testament and of Catholic cannon law. (Harari 2014)

In another area of the world, the clever minds of *Homo sapiens* found a way to survive a food shortage problem caused by population growth and lack of wild animals due to over-hunting. In the area around what we call southern Turkey, western Iran, and the Levant, they started farming. Prior to this, when graves were found, there was no evidence of farming. Graves dug revealed that only occasionally were they buried with a pet, which would indicate that they did not have the concept of domesticated animals.

Suddenly, in the agricultural revolution, about ten thousand to twelve thousand years ago, animals were domesticated, and seed crops were established. The seeds of those crops made it around the world to every continent on the planet in a few hundred years. This happened quickly in terms of history. Even today, ninety percent of the calories that we humans eat come from the domesticated plants that were developed three thousand years ago. This happened simultaneously, throughout all of civilization. Central America developed maize and beans; China had rice and pork; South America had lamas and potatoes; Guineans had sugar cane and bananas; and North America developed sunflowers and squash. By the time of 100 AD, almost every Sapiens on the planet were agriculturists and practicing animal husbandry.

[26] For detailed reading I recommend the book *Sapiens*, by Yuval Harari, along with his two other books *Homo Deus* and *21 Lessons for the 21st Century*.

Overnight, *Homo sapiens* had to work at farming every day of the week, but it was necessary for enough food to be produced. Incidentally, we have become so good at producing livestock that if you took the weight of every human being on this planet today, it would figuratively total approximately one hundred million tons. Likewise, if you took the weight of every wild living thing, including whales in the ocean, they would also figuratively weigh the same, about one hundred million tons. However, if you take the weight of every living domesticated farm animal, they would weigh about seven hundred million tons; seven hundred times as much as us, the humans who look after them.

Agriculture allowed smaller units of land to become more productive than when it was used for hunting. Overnight, land would support many more families. Now that Sapiens were living together in closer community, the social aspect of their lives increased, and a baby boom occurred. The number of people continued to increase exponentially. The measure of success of a species is how many copies of themselves are created. Agriculture enabled this to become a magnificent success.

Stop, Think, Discuss

Scholars would call the social engineering that keeps us in peaceful communion with each other social myths; myths like justice, religion, government, law, and so on. How would you explain this, in your own words, to someone else without using the term myth?

Success in agriculture also created a new era with no recourse. Once mutation has occurred, there is no going back. Explain how social change of the future will duplicate this process of no recourse.

Question one: In these statements, what is the divine intention of God? Is God even involved? How does this historical and scientific analysis blend with the religions we have been discussing? Can theology and science communicate together on these issues with validity? How?

Question two: When asked to critique a process, it is incumbent upon you to also recreate the process back to its starting point with any new information in order to support or modify the original hypothesis. From a theological perspective, what would your postulation be now that we know more about the history of *Homo sapiens* on Earth? And from a scientific perspective?

Question three: What appears to come first in this? Science or theology? Can science and theology agree on anything in this subject? What?

Question four: Absolutely everything that happens has a precursor that makes it happen. Then the event happens. Can you recount the precursor to the changes discussed here that changed the existence of *Homo sapiens* on Earth? Could it have been cataclysmic, like a volcano or comet, or more general and subjective?

Question five: How would you describe the historical changes in man in your own words?

Question six: Are there issues in these statements with which you disagree? Explain.

CHAPTER 8

Agriculture Becomes Society

Community developed. Cities, education, structure, justice, injustice, religions, and politics flourished. As Sapiens increased in number, so did infant mortality. I remember reading somewhere that apparently mothers were dying in giving birth far more often than they did when they were hunters and gatherers. Disease multiplied and evolved to become uncontrollable in some eras.

History has a rule. Luxuries, once invented, become a necessity. Once a luxury is a part of this new lifestyle, it breeds obligation, and it is taken for granted. It is misunderstood that this new luxurious life will provide a more leisurely life. But, unfortunately, this is usually not so. This new luxury requires more effort to maintain. Think of modern-day use of the cell phone and/or the Internet. Twenty years ago, they were a convenience, not compulsory for everyone to have on their person always. Technology has made them relatively inexpensive, but not entirely so. Regardless, almost everyone under age 40 has developed a sense of compulsory need to retain a cell phone with Internet connection for private use. Now there is the required need to earn extra income to support this new luxury/necessity.

And so, the era of hunters and gatherers ended with the new need for food production. Clearly this has been the same over the ten thousand years since the Agricultural Revolution started. We still have a population explosion and food production that needs constant improvement to feed all the extra mouths in this small world. So, the human quest for an easier life resulted in change that transformed our world into one never imagined or even desired.

Or was this divine intervention? In 1963, scientists discovered Göbekli Tepe, which was built around 9500 BCE in southern Turkey.[27] This is six thousand years

[27] How was Gobekli Tepe discovered? https://www.quora.com

prior to Stonehenge and runs contrary to the prevailing scholarly and historical view of the rise of civilization. When it was unearthed, it contained ten pillars, not unlike Stonehenge, weighing fifty tons, one of which was about one hundred feet across. Amazingly, Stonehenge was built in 2500 BCE by an agricultural society. Göbekli Tepe predates this by 6,000 years, during what is supposed to be the hunter-gatherer period. Clearly, it took hundreds if not thousands of people to build this monument, and it also took a long time. Food had to be provided, and wheat was found from nearby areas. Could this be a theistic or Godly intervention?

From a Smithsonian article reflecting this discussion:

> Gobekli Tepe's builders were on the verge of a major change in how they lived, thanks to an environment that held the raw materials for farming. "They had wild sheep, wild grains that could be domesticated—and the people with the potential to do it," Schmidt says. In fact, research at other sites in the region has shown that within 1,000 years of Gobekli Tepe's construction, settlers had corralled sheep, cattle, and pigs. And, at a prehistoric village just 20 miles away, geneticists found evidence of the world's oldest domesticated strains of wheat; radiocarbon dating indicates agriculture developed there around 10,500 years ago, or just five centuries after Gobekli Tepe's construction. (Smithsonian.com 2008)

After much scientific testing, it has proven that this was a pre-agriculture society. Why would a society of nomadic hunters build such a structure? It seems to have no purpose other than theism. It was not for protection, nor manufacture, nor anything else useful at the time. It is possible that these people built this area as worship, and then it created a society, which needed food. Wheat was originally gathered but then cultivated to support society. This is conjecture since no records exist to verify or deny this premise.

Along with the gathering of seeds was the domestication of animals. This has not been the best of times for the animals. Not until recent times has the domestic herd been thought of as requiring an attitude of care. Our stewardship concerns in our religions with realization that all of God's creatures have value has permeated our history of agriculture in modern times. The herds have increased to the point where it is said that domestic animals now outnumber humans by seven hundred to one.

Our human population has also increased exponentially. As hunter-gatherers, the population estimate for the entire world was five to eight million nomadic people. That number of food gathers or hunters dropped to one to two million by the end of the first century, and they were mostly located in Australia, North America, and Africa. At the same time, 250 million farmers populated the "civilized" world.

Farmers ten thousand years ago also had changes in the concept of home. They became attached to one place, one family, and one plot of land, unlike their nomadic ancestors. They protected this land and their family. They "improved" their land with canals, hedges, and fruit trees, built houses, and eradicated pests. They collected artifacts and created a home. These enclaves were small and surrounded by nature. As late as 1400 AD, farmers in this way occupied only 2% of the arable land of the Earth. Remember 70% is water.

Society became agrarian, so farmers primarily learned firsthand to worry about the future. They could not control the weather. They cleared and improved more land to provide more food for the uncertainty of the future. They grew more than they could consume and thus built up reserves. They constantly worked their fields, held momentary celebrations, and then went back to work. This stress became the basis of large political and social systems. Sadly, the diligence of the farmers never gave them the security of their efforts. Potentates, rulers, and elites rose up and lived off the labors of the farmers, leaving them with only a subsistence existence. This is just the nature of *Homo sapiens'* typical interaction: leaders and followers. Those who make decisions insist their will be followed, and those who abstain from the leadership struggle and follow.

Until recent times, ninety percent of the people were peasants, who worked the farms. The fruit of their toils fed the minority elites, kings, government officials, soldiers, priests, artists, and scholars—all the folks who create the pages of history. History is what very few had been doing while the majority were farming. After spending a month in India recently, I saw this in real life albeit on a lesser and modern scale.

Now the Sapiens have moved quickly from the hunter-gatherer stage into the building of society. Although humans did not have time to evolve, the structure of the people was enough to create a way of permanently living together. Cities were born, kingdoms arose, armies of nations were built for protection, great religions were formed, and stories were created to support these religions. Justice, government, politics, and taxes came to the forefront of everyday life.

Around 221 BCE, the Qin dynasty of China was formed, followed by Rome in the Mediterranean. The taxes levied on 40 million people in China supported an army of hundreds of thousands and about one hundred thousand officials. At the same time, Rome became the Christian Empire and collected from about 100 million individuals, which supported infrastructure such as roads and theaters that are in use to this day. Pity the poor peasant as the tax collector wiped out all the surplus food crops of his toils for a year. The only people who fared worse were slaves.

Stop, Think, Discuss

Did you know that one way to end up being a slave in ancient times was to borrow from a person and then suffer the inability to repay your loan? If you went bankrupt, you could end up becoming the property of the person from whom you borrowed the money. This gives another viewpoint to the idea of being poor, in our modern perspective.

How would you solve the problem of bankruptcy today and still have trust in democracy?

Question one: In these statements, the activity of man is observed. What do you think the intention of God is in this whole process? Is God even involved? Is this evolution of how humans change through adversity God's creation or man's?

Question two: What is the *ontology* of these statements? That is, what is the actual information, free from our personal bias, of these statements? And what is the *epistemology*? That is, what is the knowledge, as learned by us, including our preconceived notions, of these statements?

Question three: Why did *Homo sapiens* become farmers? Was it an action created by God or by man?

Question four: Absolutely everything that happens has a precursor that makes it happen. What do you think was the precursor to Sapiens becoming agronomists? Could you have determined this change in the activity of man, or the cause of the event before the change? Also, could the event have been prevented if the cause had been known beforehand?

Question five: How would you describe this transition in your own words?

Question six: Can you present a different point of view on this transition theory? If yes, then explain.

CHAPTER 9
The Inequality of Man

How Were These Empires Sustained?

In his excellent 2015 book *Sapiens*, the brilliant author Yuval Noah Harari states: "All these cities were 'imagined orders'. The social norms that sustained them were based on a belief in shared myths."

Harari gives two examples of the best-known myths in history: The Code of Hammurabi of 1776 BCE, which served as a cooperation manual for hundreds of thousands of ancient Babylonians; and, the American Declaration of Independence of 1776, which today still serves as a cooperation manual for hundreds of millions of Americans. I am going to use the thought process originally crafted by Harari here to create controversial argument. These ideas are excellent examples of providing personal bias in an argument, which allows us to contemplate which is the more desirable idea. I love controversy as a learning platform. The trick in this method is not to gravitate to one or the other side of the argument but rather accept the arguments of both in context to each other and then enter the evaluation process, often with others.

The first example used by Harari is within the area of Mesopotamia and from the Babylonian King Hammurabi in 1776 BCE. He created the Code of Hammurabi to give his subjects the persona of his being just in a uniformly applied legal sense.

Essentially the code is an eye for an eye: Equal retribution for original sin but based upon your station in life. This is where the code is very different from modern use of equality. Hammurabi breaks society down much like the caste system in the Hindu religion and even invokes the power of God into this alignment of status within our species.

His subjects are divided into two genders and three classes of people, which become the superior class, commoners, and slaves, and all are valued in silver, which was the currency of the day.

Hammurabi's justice is served when an injustice is done and the equal and opposite justifiable reaction is perpetuated upon the offender. But the equal and opposite is in terms of currency, not in equality as we know it. So, two superiors disagree, and one puts out the eye of the other, then justice demands that the eye of the offender is also removed. However, if that offender is a superior woman and the hurt party is a man, then justice might be several coins of silver. Likewise, if the offender is a commoner, the retribution is different than if he was a superior. Perhaps death? Conversely, if a superior killed a slave of another superior, the value will be several coins of silver. This was the widely acceptable law, which regulated their society nicely.

Take a time jump forward to 1776 in our common era and the Declaration of Independence of the United States of America. This declared the universal principle of justice, which was inspired by Divine influence. It states that if followed by all people within the 'kingdom', a prosperous society that cooperates each with the other will live safely within its borders. It promotes individual freedoms; equal justice for all regardless of race, religion, or culture; and the rights of everyone to participate equally within this society.

The contrast in polarity between these two systems of justice with remarkably similar dates of incorporation is as follows:

The American Declaration states all people are created equal within the boundaries of their influence. Babylonians declare that people are decidedly unequal. Both are certain that their system is correct and even blessed by God. Harari states that both are wrong. Not that one is wrong, and one is potentially correct; both are incorrect. He even states that both are governed by value in currency and even less if one is a slave. All of this is in the name of justice, but justice with hierarchy. He goes further to say these ideas of equality "exist only in the fertile imaginations of Sapiens, and the myths they invent and tell each other." Further, "they have no objective validity." (Harari, *Sapiens*, 2014)

But the controversy is just beginning. Now we get into the crux of the argument and the purpose of writing this for controversy. I give Harari full marks for giving us this opportunity to think about our role on this planet. He concludes that it is easy to conclude that dividing people into divisions, such as superiors and commoners, is a figment of imagination. "Yet the idea that people are equal is also a myth. In what sense do all humans equal one another? Is there any objective reality outside the human imagination, in which we are truly equal?" (Harari, *Sapiens*, 2014)

Still with the memory of Harari present in our minds, we evoke the now acceptable biological science of biology and accept that we were not created, but we have evolved. Further, that we accept *Homo sapiens* did not evolve equally, are not identical, and we certainly did not evolve to be equal.

Now it is up to you, dear readers. While the description of this phase of the progress of humans on this planet is good, is it completely up front and real? Does authorship give personal bias? Remember the description of ontology in philosophy. In objective science, reduce your thoughts to a basic form of thinking without personal bias and without being influenced by personal history. Some of you will find the statements about creation unrealistic and anathema to your personal belief of creation. Others will agree with Harari, and yet others will respect the history and ascertain that Harari has implanted his own bias into his telling of history. Perhaps yet others will find an alternate equation to making the creation story fit the facts of history.

> *If you would be a real seeker after truth, it is necessary that at least once in your life you doubt, as far as possible, all things.*

René Descartes[28]Language and Culture

Language is the commonality in the structure of large groups and their belief systems. This language is used to maintain and convey culture and cultural ties; it becomes the bond that unifies a large group. The whole intertwining of these relationships starts at one's birth.

Language is not just used as a tool for the exchange of information, but as a system with the power to create and shape concepts, such as values or ethics, perceptions, identities, and culture, and to share human values within a social group. It includes the study of philosophy, religion, history, literature, and the arts. The Sapir-Whorf hypothesis[29] indicates that if you learn another person's language, you learn how he

[28] Rene Descartes Quotes, Good Reads, https://www.goodreads.com
[29] Linguistic relativity, sometimes called the Sapir–Whorf hypothesis, is a hypothesis in linguistics and cognitive science that holds that the structure of a language affects its speakers' worldview or cognition. The strong

thinks. I'll use my travels to Japan an as example. They, through their language seem, to not have the ability to say "no". If their response to a statement is negative, they will go into several layers of how this cannot be done, or it is not possible, but a direct "no" is almost never stated. I imagine that this is required in their culture because of the potential "loss of face". Loss of face is another whole dimension of their culture that I will avoid for now as I personally got it wrong by giving a valuable gift to someone (a restaurant owner), who was not expecting it and therefore could not offer a similar gift in immediate return. In the ninth century AD, data became so cumbersome to remember that another script was created, initially by the Hindu-speaking people, but it was quickly taken over by the Arabic nations that overran the Hindus. Numerical script was created. It also evolved with mathematical signs for addition, subtraction, multiplication, and so on. This has become the world's predominant language, conveying ideas with graphic clarity to everyone who reads it, regardless of their written or spoken language.

History is about the making of a country. Culture is about the making of an individual. But both are interrelated, too; culture is a subset of history. Most often, scholarly focus is on experiences shared by "peasant" groups in a society, like art, music, and dance. Where history is about the past, culture is about the mix of the past and present. A culture of a land will become a part of the history of that land, and a history of a land can be due to the richness of the land's culture. Hence, you can say that culture is a subset of history.

Culture is heavily influenced by the religion of that land, and sometimes that is also a reversal of process. Did the religion derive its manifest from adapting to a culture, or did culture form as a derivative of religion? How do you reconcile the idea of not eating pork or wearing a headscarf in public? What is the basis of these beliefs?

Then there is the culture of the hierarchy of the sexes. In ancient history, culture allows that in many societies, women are property, and that an unmarried woman can be raped with all the consequence of finding a coin on the street. The law in some societies indicates that rape of a married woman is not a violation of the woman but rather a crime against the man to whom she is married. So, the question becomes: Is this division of men and women a result of imagination, like the caste system in

---

version claims that language determines thought and that linguistic categories limit and determine cognitive categories. en.wikipedia.org

India, or is it a result of deep biological roots? The biological rule is that biology allows whereas culture forbids. Biology allows a broad spectrum of possibilities, while making others impossible, such as giving birth is restricted to women, while having a sexual partner is not. Therefore, culture determines the outcome in societal acceptance of each of these actions. For example, it is culture (often influenced by theological sources) that rejects or accepts gay marriage, and culture that declares the equality or inequality of women. Biology is like a mathematical equation. It is either possible to do or not; no emotional values or subjective reasoning influence the determination.

Another theory is that masculine dominance stems not from strength but from aggression. As a male, I have issues with the idea that evolution has made men far more aggressive and violent than women (which is often represented by legal statutes and politically biased groups). Women are men's equals in hatred, greed, and abuse, but men are far more willing to engage in raw physical violence. Throughout history, war has been a masculine undertaking.

Stop, Think, Discuss

Finally, there is the big question of philosophy. We Sapiens have historically centered our culture, our thinking, and our understanding around the self in all things. Historically, we believed we are at the top of hierarchy in the animal kingdom, on the planet, and in the universe. In the fashion of philosopher René Descartes, ask yourself:

+ If Sapiens did not exist, would there be a universe? Would it matter?
+ In the big picture, we are insignificant, but do we understand this?
+ Perhaps even ask yourself: What is it that we do not even see, do not even know exists, and are we necessary for any of this in the first place?

Some of this thinking also exists in part four here, about the stars, concerning dark matter and dark energy, but we do not have the smallest clue as to what it is.

Question one: In these statements, we now realize that we are all related one to the other. So how do we differ if we are all the same species?

Question two: What is the *ontology* of these statements? That is, what is the actual information, free from our personal bias, of these statements? And what is the *epistemology*? That is, what is the knowledge, as learned by us, including our preconceived notions, of these statements? Use the Harrari quotations to discuss.

Question three: Culture also creates language. How does theology or the different religions affect culture and has this made us unique tribes? Can we also be the same even though we have cultures that are different?

Question four: Absolutely everything that happens has a precursor that makes it happen. If we are all the same with the same chromosomes, what happened to make us different in our own minds?

Question five: How would you describe the differences of language and culture and their creation in your own words?

Question six: Are there issues in these statements with which you disagree?

CHAPTER 10

Government and Finance

Politics and the Economy

All philosophers agree that in the review of our cultural history, the only thing that stays the same is the act of change itself: change in the deity of different gods and change in the structure of society from tribes, the reigns of kingdoms, oligarchies, dictatorships, spiritual or religious zeal, "gangsterocracy,"[30] tyranny, and democracy. Even these have sub-groups with degrees of good and evil. For instance, dictator Josip Broz Tito in Yugoslavia was able to keep the country unified, and there is Lee Kuan Yew's benevolent dictatorship of Singapore, where the citizens love their leader. Kings through the ages have been everything from benevolent to cruel. Sharia law keeps the Islamic world of ISIS under the leadership of a few imams who, perceived as the ultimate authority, revert to ancient texts of their teachings to invoke state law and order.

Tribal leaders were the head of a family or the strongest member of a group. They could create the rules and designate who or what could have sway within the tribe. Any members of the group who had ideas counter to the leader were removed, either by dismissal or violence. Might is right. Gradually tribal reign gave way to kingdoms, where the leader not only ruled by might but also by divine right, with the blessing of the gods. Culture and religion played a large role in this as it evolved into reality. This evolution was in lockstep with *Homo sapiens* moving from hunter-gatherers

---

[30] *New York Times*, https://www.nytimes.com/ September 5, 2007. Iraq is a tribal society. It can only be pacified by lining up "organic local actors—some thuggish, some not". Saddam, the *capo di tutti capi,* or boss of all bosses, well understood this, preferring gangsterocracy to Islamic theocracy and enforcing his rule with a combination of a velvet glove and a brutal fist.

to agriculture. Settlements became the norm, requiring protection from other tribe settlements, and the growth industry of tribal leaders became the royalty of the new agrarian society

The emperors of Rome ruled by divine right throughout all the Eastern Mediterranean. Then, when Emperor Constantine adopted "The Way" or Christianity about 315 AD, the Christian faith morphed with this hierarchy of divine rights. God anointed the pope to rule alongside the emperor, and neither were subject to rational man-made laws. Although the Christian movement had already started to structure itself with bishops and leaders, now they took on the leadership role by divine right. This gave rise to religious leadership with powers of great magnitude. They had the power of life and death, taxes, imprisonment, and so on. This was an extension of the power of Rome. The leaders of the Church were forced into celibacy to ensure that the Church's assets were not claimed by heirs of the pope or priests.

Seven hundred years later, Muhammad created the structure of Islam. This religion also created power through religious leaders who interpreted Islam as the legal power of the land. They used the three divine books, originally dictated by Muhammad, and Sharia law to exercise power over the masses of peasants whom they ruled. This was not divine power like Rome and Christian; it was law by the will of scripture dictated to Muhammad by the Angel Gabriel.

Tribal leaders, kings, dictators, despots, and armies were the rule of the day. The French Revolution was the big change in history, whereby the common man finally realized some control over his life. They wanted equality and individual freedom, which they briefly won through revolution. These two freedoms have become the worldview of fundamental freedoms today. Unfortunately, freedom shortchanges equality, as giving someone freedom to do as they please will impact the equality of the other person with whom they are doing as they please. If a wealthy person is given the freedom to own and occupy the shoreline of a beach in Hawaii and restrict the rest of the inhabitants from using that beach, the freely acquired property of one restricts the ability of others to a nice Sunday picnic on the beach, and so begins the struggle of modern society to come to grips with this anomaly.

Democracy was first discussed in ancient Greece by Socrates and later by his student Plato, around 350 BCE. However, they were looking at democracy in a different context than modern perspective. They were trying to find a way for the various city settlements throughout Greece to come to common and acceptable agreements. Their writings indicate that they proposed a vote of each city at a gathering of the whole should result in the votes of the majority of the cities having sway. This is a leadership discussion, not individual right.

Much later, in 1215 AD, the *Magna Carta* also tried to bring a form of democracy to England. King John had lost most of his support from his nobles and barons at home after his poor oversight and losses of military battles away. To salvage his power, he agreed to peace talks with his potentates of the feudal system. They agreed to sixty-three clauses of sharing power and the king giving up some authority. King John clearly had no intention of honoring this agreement, and before the end of 1215, conflict arose again. Other countries in Europe tried to give some semblance of rights to everyone, mostly without success. They principally dealt with groups rather than individual expression within a community. Finally, England created the first parliament in the first real attempt to give the common man some power of control over his wellbeing, but still lacking in individual expression.

The American Constitution, revised in 1788, gave rise to modern democracy as seen in the world today. Individual rights were entrenched, trust was enshrined, and people were granted the ability to worship, select representative government, and live within the country with the ability to express themselves freely, without fear of retribution.

Democracy is different from all other forms of government. It would be easy to break the world into two types of governing: democratic and everyone else. Simplification would further call these two groups democratic and tribal, with tribal being a conglomeration or oversimplification of every other type of rule that exists with one powerful entity controlling at the top.

Precursors of Twentieth-Century Human Rights Documents

Documents asserting individual rights, such as the *Magna Carta* (1215), the *English Bill of Rights* (1689), the *French Declaration of the Rights of Man and of the Citizen* (1789), and the *US Constitution and Bill of Rights* (1791) are the written precursors to many of today's human rights documents. These documents reflected the culture of that time in history, much of which today is not acceptable, in our human dynamics.

Often, they excluded the rights of women to exercise their franchise to vote or even hold property, also perhaps people of color, and disenfranchised social, religious, political, and economic groups. Regardless, the basic principles of the documents drew the oppressed into support for their perceived rights through revolution.

Democratic thinking is at the very least due to the evolution of Christian leadership eventually allowing great thinkers to express their ideas without the threat of excommunication, expulsion, or death. Of course, this was not always the case in Christian history, but the result of the evolution of Christian culture over time. The equality of the individual within the society is expressly a Christian value. Love, forgiveness, and grace are some of the primary values of Christianity, and they give rise to trust and faith. At the same time, Islam requires that everyone of that faith adhere to Sharia law, which is all-encompassing in social, political, and legal terms, and mandated by someone at the head of the organization.

Democracy is self-righting, like a boat in rough seas. It has within it the checks and balances to correct itself from errors and wrongs. Everything is based upon trust. If a politician makes a mistake, an error in judgment, or worse, and it is against the will of the people, that politician will find themselves without a job after the next election. Freedom of the press dictates that unscrupulous action within a democratic government is excised through the press and made known to all. Even good government that loses touch with the common voters can lose perceived trust and find themselves outside looking in after the next election.

We now live in interesting times. Great Britain and the United States have closed their doors to the equality of man if that man desires to immigrate to their individual countries. The United States has even elected a president who seems cavalier about his historical respect for women and colored citizens of the US, and who demeans Mexicans and others, particularly Islamic peoples desirous of being a part of the once melting pot of democracy. Liberalism is dying and perhaps democracy behind it. Where is the equality of man in the United States' current state of politics? Cultural differences are the order of the day. In the United States, there are even movements that parallel the fanaticism of Nazi Germany in 1936.

I often make the argument that Christianity is the precursor to democracy. Let's look at the democratic premise of Christianity, which is in three components: love, forgiveness, and grace. Add to that trust, equality, and charity, then use all six as ingredients to form democracy. Love is the love of family, God, the land, neighbors, and all living things. Forgiveness is not only for those being forgiven but also to relieve the burden placed upon you by the actions of another. And grace is the knowledge that you are cared for.

My analysis is thus: to equate love, democracy uses love as its central core value; care for each other, for the land, and for democratic principles of equality and justice.

Forgiveness in democracy recognizes that when a person does something against the rules, then after they take ownership of their mistake (which may involve serving a prison sentence), they are forgiven and start all over with a new possibility of a good life.

A democratic system, using grace as a guide, cares for its subjects through health care, social assistance for the poor, fairness in law, and so on. It is further supported by the Christian value of equality, with one vote per person within society, wherein all votes are equal in the counting of ballots. It also requires trust from every individual and business within that government. That trust is also governed by laws and (even when necessary) with armed force.

> *A little philosophy inclineth man's mind to atheism; but depth in philosophy bringeth men's minds about to religion*
>
> *Sir Francis Bacon*[31]

Therefore, the strength of democracy is trust. Trust is also primary in the coexistent partner of democracy: capitalism.

Capitalism, Money, and Resources

Until democracy came along, the world existed with little trust in a financial sense. The financial resources of the entire world were limited to the existing resources on the planet. These resources neither grew nor dissipated. They were like a pie that got cut up into many slices, with some entities getting a bigger slice than others. There was only one pie and many mouths to feed.

The peasants saw this as unjust and were riled when they saw individuals or organizations taking more than a modest share of their pie. One of the old metaphors

[31] Sir Frances Bacon (1561–1626) was an English philosopher and statesman who served as Attorney General and as Lord Chancellor of England. His works are credited with developing the scientific method and remained influential through the scientific revolution. https://en.wikipedia.org

from Christian writings is that it would be easier for a camel to get through the eye of a needle than for a rich man to get to heaven. The basis of this metaphor is that the pie is constant, and a rich man gets more than his share. From my travels to Jerusalem, I also heard another speculation that the eye of the needle is really the gate in Jerusalem that is called the "eye of the needle".

Under this ancient system, people were reluctant to lend funds to start a business or for seed for their crops. There was no trust on either side. There was no bankruptcy, and if you did not repay your indebtedness, you would become a slave to the lender. He would own you until you paid back your debt to him. In modern thinking, that makes for one level below poor, and not a very good option, either. It also makes borrowing money a rather large life risk.

Democracy changed all that. With the creation of corporations and banking systems because of the Industrial Age, the resources of the world were not finite anymore. There could be several pies to be divided up among everyone. More pies were created every day. Suddenly there was trust between lenders and borrowers. The lender would trust that the borrower would pay him back with interest. And it worked. There had to be a symbiotic relationship between the capitalist system and democracy. Because private property, which is paramount to any idea about capitalism, grants individual freedom and gives individual rights, private property reduces governmental power and gives everyone some protection from intimidation. Further, well-defined property rights give alternatives for governance, making the government ineffective or absent in some aspects of daily living. Trust in a capitalist, democratic society fosters self-reliance and personal responsibility.

If capitalism did not have any way to enforce the return of funds should the lender renege on his obligations, it would fail. The lender would not be allowed to have his own army or court to enforce the obligation. So, under democracy, the lender would go to the courts of the democracy requesting enforcement of the rules of the agreement. He might even go to the government if the borrower was foreign and had no intention of repaying the funds. The democracy would then consider using their army to enforce the rule of law. This worked for both the borrower and the lender.

Capitalism, in this form, worked very well for a long time. Then as corporations grew, some became monopolies. Monopolies are the breakdown of capitalism, the point at which capitalism, and thus democracy, ceases to work. A monopoly exists when the corporation, through good business practice, eliminates the competition. The owners are still interested in rising dividends, wanting more and more returns for their investment. The sales are now fixed in a saturated market, so the pressure is on the owners to find ways of increasing dividends through efficiencies like taking wages and benefits from their employees and increasing hours of work. An unfair distribution of wealth occurs, and we go full circle back to peasants and potentates. The potentates influence government and the cycle worsens.

A perfect recent example is the United States. According to recent data in February 2018, at the end of 2017, fully 40% of their yearly Gross National Product (GNP) ends up in the hands of less than one percent of the population. Since then, that same one percent argued successfully with the American government for reduction in the taxes that would affect them or their businesses. That means that 60% of the GNP per year is split between 99% of the remaining population. Then look at the bottom half of wage earners, who split approximately 13% or less of the country's wealth, making them peasants by any standard of living criteria, with no hope of ever recovering. Then along came a nearly fascist presidential candidate who is also one of the one percent, but states, in his truth, that there will be a better life for them—and he gets elected. Like most egocentric *Homo sapiens*, his truth was false, and now they are a country on the decision-making path of whether they will continue to be the world's leadership in liberty, freedom, human rights, and democracy.

In some instances, monopolies can also induce in employees an attitude that causes them to feel much more entitled than society sees them. Keeping their interpersonal relationships within the monopoly, they see themselves as vital and controlling in the industry. Thus, they request much more recognition in compensation and benefits and fewer hours of work. Monopolies eventually cause trauma and sow distrust in a democracy.

The obvious path is for government to legislate or disallow monopolies from forming. Often it is a government itself that chooses to create government monopolies within their jurisdiction. Through a mild form of narcissistic behavior, necessary for anyone who wants to become a politician, they often view themselves as benevolent and much more able to control business than industry can. The result is a difficult situation whereby the government must extract funds from the government's monopoly to support a tax burden. The employees see themselves as mandatory in the running of the business, if not in the government itself. Thus, an elitist opinion of themselves as employees is created. Further, in a government monopoly, the government has the power to regulate the product, the safety regulations, the pertinent accounting principles, the tariffs, and the quality in the marketplace, thereby supporting the sale of their products and eliminating competition by regulation.

There is one other interesting aspect of a democracy that is worth discussing. Every human being has an opinion. In other forms of government, they often find it advantageous to not express that opinion for health reasons. In a democracy, they are held harmless in this because of the trust factor. Opinions are usually stated without fear of retribution.

Stop, Think, Discuss

I inject some controversy to peak your thinking.

Much like the earlier discussion on feudalism and leaders and followers, there appears to be two kinds of Sapiens in this chapter. One kind has and expresses personal opinions freely and passes judgments quickly, often without much regard as to the overall impact of these opinions upon society. This derivative of evolution only wants to hear their own ideas returned, like an echo chamber. They love their own ideas, which usually only impact others, and they want their own thoughts to be the only policies of that party. Frequently, their great desire is to have their issues financially sustained by those who are opposed and impacted. Thus, they hold themselves from any burden created by their own ideas. I see this personality type in all political spectrums of government. Other politically active Sapiens, who may even hold similar views, draw a conclusion and get on with their lives outside of activism. Could this second group be called the silent majority?

What would you say to these ideas?

Question one: This discussion is about politics, with some statements including God for comparison. Is there a comparison to God in this discussion? Is God even involved? Are politics restricted to only input by *Homo sapiens*? Does theology enter politics? Think about the theocracy of some countries. Are politics God's cultural influence or man's?

Question two: What is the truth within this discussion? That is, what is the actual information, free from our personal bias, of these political statements? And what is the preconceived notions, of these statements?

Question three: What process causes political activity within humans and is the original reason for politics in the first place? Is it created by man or God?

Question four: Why did all the political systems in the history of the world fail, except democracy? Is democracy failing now? If yes, what is replacing it?

Question five: How would you describe any of the controversial statements in your own words? How do you think distribution of wealth affects voting outcomes?

Question six: Are there issues in these statements with which you disagree? Has creative thinking replaced rational thought?

CHAPTER 11

Industry and Science Change Everything, Again

The Scientific Age and the Industrial Revolution

The dominance of *Homo sapiens* started with the Age of Cognition eighty thousand years ago. Sapiens moved out of Africa and soon dominated the world as hunters and gatherers. Then, ten thousand years ago, the need for food became dominant, and the age of agriculture started through necessity. With this, Sapiens created culture, politics, justice, religion, government, and status, while still operating within a family unit. Then, the predominant change occurred five hundred years ago.

The Scientific Revolution supersedes all the changes in Sapiens' development prior to this period. Medieval Christianity was the dominant political culture around the Mediterranean, Europe, and North America. The Scientific Revolution was the purview of the Christian masses, thinking outside the parameters of their Christian masters who had previously controlled their destiny. While Islam and Hinduism influenced Asia and Africa, they had mostly given up on scientific curiosity and philosophic thought due to the instruction of their leaders in 1000 AD, who insisted they concentrate on theology in their studies. The Scientific Revolution preempts the earlier Renaissance and Reformation, as the Renaissance only internally shifts within the system of medieval Christianity. Christianity had become the scientific reasoning community, while Islam moved away from their leadership role in scientific reasoning.

Almost universally, scientific reformers were priests educated within the Christian theistic system. They were instructed by the Christian Church, such as Saints Albertus and Thomas Aquinas, to be part of a system that desperately needed science to confirm the writings of the Bible. Christian leadership was still autocratic in its teachings so there was to be no possibility of the sacred texts being metaphors for an ancient people who were uneducated and unable to read or write. The leading authorities within this theistic system wanted everything to be interpreted as literal, easy to teach, easy to reinforce, and easy to defend. In history, to accomplish this, the system had previously destroyed schools of learning as far back as Greek philosophers when it was decided by the power of man that some of the Greek philosophy was not in agreement with Christian interpretation of the Bible. Regardless, these scientific priests with academic scientific reasoning and visible fact pursued their endeavors to develop the foundations of our current scientific thinking—a scientific process that has always been changing and moving forward, even to this current day. As soon as we think we have heard of a discovery that seems to make sense, other brilliant scientists move the boundaries just a little. Consistency is the playground of dull minds. As we have learned from our history, as soon as someone tells you they have all the answers, move quickly away from their thinking because it will change very quickly.

> The learned men of former ages employed a great part of their time and thoughts searching out the hidden causes of distemper, were curious in imagining the secret workmanship of nature and ... putting all these fancies together, fashioned to themselves systems and hypothesis (that) diverted their enquiries from the true and advantageous knowledge of things.

John Locke (1632–1704)[32]

Visionaries as Heretics

Visionaries of this era were often branded heretics, disgraced, burned at the stake, or imprisoned until they recanted their discoveries.

Such visionaries included:

+ Copernicus and Galileo for stating that the Earth revolved around the sun;

---

[32] https://www.iep.utm.edu John Locke (1632–1704) was among the most famous philosophers and political theorists of the 17th century. He is often regarded as the founder of a school of thought known as British Empiricism, and he made foundational contributions to modern theories of limited, liberal government.

+ Martin Luther and John Calvin for daring to question the authority of the Church;
+ René Descartes, the philosopher who stated, "I think therefore I am";
+ Kepler, who discovered that the orbit of Mars was elliptical;
+ Francis Bacon, called the father of empiricism;
+ Spinoza, along with Descartes and Locke, who were leading philosophical figures;
+ John Milton, who wrote about defending free speech and freedom of the press;
+ Voltaire, considered the father of the French Revolution; and,
+ Charles Robert Darwin for his work on the evolution of species.

Three hundred years after scientific discoveries permeated Western thinking, the Industrial Revolution began. First was the change of attitude of commerce, as previously discussed. Second was power to make machinery work, and the third was written language for communication, which spread thinking throughout the whole industrial region. The heart of the Industrial Revolution was the conversion of energy. Coal was initially the way to make steam, which powered early machines, and subsequently made electricity. Enormous amounts of coal and ore were necessary, and that became a resource that was coveted by other countries, along with the other four reasons we were taught in school as to the cause of the First World War. Remember? MAIN—militarism, alliances, imperialism, and nationalism. Militarism meant that each country was conscripting larger and larger armies. Alliances between countries created a problem when two of the opposing alliances had a problem with each other. Imperialism was problematic because European countries were gaining colonies and others wanted to join in on collecting, thus building national wealth. Finally, there was nationalism, or superiority complex, and a desire to create one's own country away from the Hapsburg rule. All of these led to the First World War. The abysmal resolution of the First World War led to the Second World War because of reparations required from the conquered nations. Some would say that another world war is not possible because the resources of the world are now knowledge and technology, which can move quickly if in jeopardy.

The Industrial Revolution needed labor, and paid wages for individuals to leave family farms to work the business of industry. This was the beginning of the breakdown of the family unit. Workers left their extended families on the farms and moved to cities to work and earn a living without the support of their whole family, leaving the family still on the farm without support. It also introduced these new workers into the field of consumers. They wanted things to replace the family comfort and nurturing that they were used to and began using desirable objects to console themselves. They needed shelter within the city where they worked, and this created another skill of the working class, building shelter and industrial buildings. Suddenly, the world was changing. Cities were growing to enormous sizes, transportation became necessary beyond the horse and buggy, and even the long-held desire of man to fly became a reality by the end of the century.

This was no friend of the average worker. Bankers and capitalists became rich, while the laborer was impoverished. Abusive labor practices became even worse as Third World countries were exploited. The slave trade was initiated and condoned by the governments that benefited from this practice. Capitalist greed reigned supreme. Although the economic pie was much bigger than it had been at the start of the Scientific Age three hundred years earlier, the distribution of wealth had become concentrated in the hands of very few. For a time, a new system of government called communism tried to change this. It failed because it was even worse than the capitalist system it tried to replace. However, communism introduced a socialist arm to the capitalist system, one that was attempting to create equality within the capitalist system mostly via redistribution of some of the wealth to those with less.

The socialist variation brought universal health care to some regions, universal education, and supports for the disabled and less fortunate. It also introduced variable taxation of the population, based upon income earned. This, along with democratic representation, creates an opportunity for even distribution of wealth among everyone. The measurement of the success of this is a better life expectancy, less child mortality, and improved nutrition. Life in most societies in 2019 is decidedly better than it was one hundred years ago.

The breakdown of the family started with the advent of the Industrial Revolution and the mass movement to the cities to provide labor for the machinery of industry. Then, in the 1960s, the birth control pill was created. Perhaps future generations will see this as the age of the pill. This small chemical wonder allowed female emancipation. No longer were women burdened with the labor of raising children and cleaning and preparing food for their families. They were able to educate themselves in the same manner as men and were now able to join the labor force, which carried prior conceptions of lost time from work due to child rearing and homemaker status. Evolution has not been as quick to move, and our DNA still longs for the nurturing and loving mother that historically has been our way of life, even though vast numbers of mothers generally work full time and children are with nannies or daycare.

The female has joined the male to become more sexually active without marriage since the worry of unplanned parenthood has substantially diminished. Marriage itself is not in vogue with many cohabitating adults anymore. They can have all the social interaction of marriage without the "old fashioned" female burden of marriage. As attitudes change, men have started to become more introverted in the presence of females. Previously men-only roles are now understood to be open to both men and women. This role change has been both good and bad for modern society. Integration of the sexes has made for a much healthier work experience, a greater understanding of others' needs, and a desexualizing of workplace role models.

There is also a definite downside to this experience. Birth rates are down in Western society. If one looks at statistics, the decline of birth rates has an equal and opposite graph in the increase of the ownership of dogs. Sapiens in economically advantaged countries have become a dog society. Dogs represent unconditional love, and our DNA, which has not evolved as quickly, still needs this unconditional love. Social diseases, STDs, have increased and even mutated into new varieties that are not treatable. Since the capitalistic societies in which we live require labor to maintain a socially acceptable lifestyle, the birth rate in developed Western countries has dropped below that required to continue the population levels at a level pace. Immigration from Third World cultures has become the norm in economically successful countries. This is changing the face of existing culture in Europe and North America. A birth rate of 2.1 children per female is required to sustain the population at a constant level. Here are the statistics of birth rate from 1960 to 2014 for the countries listed.

Table 1: Birth rate per female

Location	1960	2014
The world	5.0	2.5
The EU	2.6	1.5

OECD	3.2	1.7
North America	3.7	1.8
Latin American and the Caribbean	6.0	2.1
The Arab world	6.9	3.4
Sub-Saharan Africa	6.6	5.0
Least-Developed Countries	6.7	4.2

Note the birth rate in African countries, where they need a much higher birth rate to compensate for the AIDS epidemic.

Inability to See into the Future

The human failing of not being able to foresee the future also hints at the possibility of exhausting all resources resident on this planet. If history were the basis of judgment, this would be true. Most animal life has been reduced to domestic animals, and for food only. However, scientific human endeavor has consistently overcome these forecasts. As example, one hundred years ago, lumber was the predominant construction product, with brass, copper, and tin as dominate industrial materials. When it became necessary, steel, iron, titanium, and aluminum were put into everyday use. This could not have been predicted in 1900. Because we cannot see into the future, we cannot see how our use of raw materials will change over time, nor how it will affect Sapiens.

Not being able to see into the future has also created an almost religious zeal in the precept of contamination. A population explosion has produced a byproduct of waste, pollution, and an unbalanced makeup of the gases that are the envelope of air necessary to support life on this planet. Ideas and potential solutions abound around this subject. All are purporting to be supported by the scientific community, which has taken on a god-like status. Just like the ancients, environmental theory is centrist on the Sapiens species and does not recognize influence from any other factors or alternate scientific processes.

Another interesting evolution for Sapiens is an acceptance of scientific values over historical theological values. Mostly, it has been a Christian culture that has generated scientific progress, at least since approximately 1500. Scientific values have taken modern day prominence, likely because of ancient Christian theologians' resistance to scientific discovery as it related to the biblical story written in ancient texts. Theologians of old insisted on their biblical interpretation of the creation of babies within the womb to approach what would be witchcraft today. Evolution, as disclosed by the research of Darwin and the resistance of the theologians, is another example. Other religions have also resisted evolution in their thinking by becoming fundamentalist and dogmatic in their version of faith.

Science without religion is lame, religion without science is blind.

Albert Einstein (1879–1955)[33]

Most Sapiens today would indicate that they are spiritual and/or politically correct. Maybe they don't want to dispose of religion just yet, just in case? But their stated opinion is considered an evolution toward a reasoned, science-based, proof of perspective. Sadly, it is also without much deeper meaning and investigation into human philosophy and natural human tendencies of love and faith, both of which form deeply meaningful values in human construct.

Few men think. Yet all have opinions. Hence, man's opinions are superficial and confused.

John Locke (1632–1704)

The current myth of Sapiens is to be politically correct today and rectify yesterday's thoughts, which were in context back then. Unfortunately, this has become a process of retribution for many of those supposedly seeking solace or to correct past actions, which were thought to be correct at that time. This is not productive for living in harmony as a species. We need to forgive, forget, and find a way to the future together, in harmony with each other. We evolve and hopefully in a good way. Only by learning to live in harmony by caring for each other, forgiving, and learning that we are valued and cared for within society will Sapiens continue to exist in peace amicably.

---

[33] http://www.planet-science.com Albert Einstein (1879-1955), The father of modern physics

Stop, Think, Discuss

When we discuss community, trust is the binding that keeps our current system intact. The word "integrity" also is derived from and part of this trust factor. In the whole world, there seem to be two ways to view integrity. One is a civic or public personality whereby one has trust within the community and always tries to do the "right" thing. The other has the same trust value but is more in keeping with the family, whereby anything can be done outside the family unit to keep the integrity of the family.

What is your definition and understanding of integrity?

Question one: In these statements, I made the controversial observation that the birth control pill is a significant precursor of change to human activity. Would you agree or disagree with these observations?

Question two: What is the *ontology* of these statements? That is, what is the actual information, free from our personal bias, of these statements? And what is the *epistemology*? That is, what is the knowledge, as learned by us, including our preconceived notions, of these statements?

Question three: Is there a possibility that the Industrial Revolution had anything to do with God? Was the change in civilization created by God or by man?

Question four: Absolutely everything that happens has a precursor that makes it happen. Then the event happens. What was the precursor to the Industrial Revolution?

Question five: How was the Industrial Revolution tied to the Reformation?

Question six: Provide a statement where you agree and/or disagree with the thinking in this chapter.

PART THREE

Evolving, Biochemistry, and DNA

*To learn, one must not only read but must also think.
Thinking is best done through discussion.*

Chris Pedersen

Stop anywhere it would be prudent to ask any question.

CHAPTER 12
Evolving and Biology

Homo sapiens evolved from life generated on planet Earth millions of years ago. More than one species of upright-walking beings exited the dinosaur ages. In part two, we talked about the four upright-walking species that existed eighty thousand years ago: not apes but men, people with brain cavities larger than apes. Today, all humans in the world are directly descended from the *Homo sapiens* species that was originally of Africa, but which migrated to every part of the planet. In the process, *Homo sapiens* mingled with Neanderthals and the other two species; thus, there is measurable Neanderthal DNA in our genome.

However, *Homo sapiens* were not even at the top of the predatory chain in our history. Headed by Sylvain Bonhommeau, Laurent Dubroca, Olivier Le Pape, Julien Barde, David M. Kaplan, Emmanuel Chassot, and Anne-Elise Nieblas, a French organization looked at the human position on the predatory scale, called the trophic level, in 2013. They combined ecological theory, demography, and socio-economics to calculate the human trophic level (HTL) and position humans in the context of the predatory food web and published their results. The unflattering conclusion they reached, through this HTL calculation was that humans aren't all that superior. We are right in the middle of the food chain. This is reinforced by a newspaper article I recently read that called humans the bone crushers, which describes an animal that feeds after the higher forms of life have had their fill of a kill. They crush the bones of what is left for the nourishment of the marrow, which effectively places human's way down in the middle of the predatory (or food) scale.

For those Christians of us who, in our own minds, often considered ourselves as founders of scientific reason, this is a shock, especially for those who want to believe in a literalist interpretation of the Bible. We have arrived at a stage in the development of *Homo sapiens*, called cognitive dominance, which involves efforts to impose or maintain a pointed set of ideas or beliefs of one group onto other groups, often using many types of mentally enslaving techniques, i.e. cognitive dominance. The first book of the Christian Bible, Genesis, which is also a foundation of Judaism, has two stories regarding the introduction of man and woman onto Earth. Read Genesis 1:1–31 and compare that to Genesis 2:1–25. Both are a metaphorical story to help the ancients understand a principle of living that needed to be understood but which the ancients were incapable of grasping without a story that made sense to them. Remember, very few were even able to read or write, and education levels were not high. They historically lacked the education or the ability to reason beyond what they saw in the present. These Genesis stories were invented to convey a principle of living that helped ancient people create a life of grace. These and many other myths were also the psychological foundation for growing a society beyond the family unit.

Psychologist Leon Festinger proposed a theory of cognitive dissonance centered on how people try to reach internal consistency. He suggested that people have an inner need to ensure that their beliefs and behaviors are consistent. Inconsistent or conflicting beliefs lead to disagreement, which people strive to avoid.

Another interesting effect that I have been alluding to for years is the Dunning-Kruger Effect. Named for Cornell psychologist David Dunning and his grad student Justin Kruger, this is the phenomena wherein people who are unskilled or ignorant in a domain believe they are much more competent than they are. Thus, volunteers in a nonprofit business who have no business skills in that field believe they have good business acumen. Incompetent or untrained individuals (not feebleminded) believe they are as able as trained people on volunteer boards. Dunning and Kruger documented this effect in several ways. The authors observed that you need skill and knowledge to judge how skilled and knowledgeable you are yourself.

Then jumping to other religions, we have other vastly different stories of the creation of man all without scientific reason. Ask yourself: Are all theologies disregarding the age of dinosaurs, all purporting to be a gift from God. Are they all metaphors giving humans something to think about or giving a starting point at the time of writing these creation stories?

It was necessary for man to attempt to understand how we commenced our historical lineage. However, the genealogy of *Homo sapiens*, as evolved from life created on Earth, cannot be told to this day as it is not understood. The root question is this: What is the cause that resulted in the evolved development of man? The origin or formative effect of our species is not known; only the result, which is us. This allows speculation as to the origin of *Homo sapiens*, providing room for a creator or theistic thinking. Think of the immortal cartoon of Pogo in the swamp following his own footsteps and concluding, "I have found the enemy and he is us."

Seeking Pleasure

Millions of years of evolution have generated a psychology of biological pleasure that has become the motivational factor in the development of our species. A euphoric upper that follows successful completion of an arduous task or the solving of a complicated thought process are examples. Fear is also a motivational factor, but it is a negative one because it causes paralysis. Pleasure, or more specifically, the sudden biological enhancement or uplifting of pleasure has been sought throughout history. Simplistically, I refer to this as the pleasure center, which is the nucleus accumbens of the brain.

This so-called "pleasure center" was co-discovered in 1954 by Peter Milner while he was a postdoctoral fellow at McGill University and James Olds, who was an American psychologist. They stumbled on the pleasure center after they implanted electrodes into the brain of a rat and found that it became addicted to pushing a lever that stimulated the nucleus accumbens by adding dopamine.

In 2014, the findings of Dr. Jonathan Britt, from McGill University, used a new brain imaging technique to determine specific inputs linked to pleasure and addiction coming from different regions of the brain. He found different ways that this same part of the brain uses dopamine-dependent reinforcement signals with environmental stimuli; regardless of the constant state of every day, an enhancement of our sensations is always sought. For example, today, a person living in high status, expending unlimited funds with seemingly unending happiness, seeks the sudden burst of enhanced biological pleasure to bring personal satisfaction. The very same motivation is there for the opposite person who lives an existence of poverty and lacks accommodation. The burst of satisfaction, or pleasure, is often the result of the accomplishment or success of the task at hand being successfully completed.

Other things have been used over the years to enhance our senses and provide pleasure via dopamine in the nucleus accumbens. *Homo sapiens*, during the age of agriculture, eighty thousand years ago, discovered that the fermenting of grains and fruits produced a beverage that, when ingested, produced an intoxicating experience. We have used this phenomenon ever since to periodically enhance our pleasure center. Our reproductive biology gave us pleasure in the venereal act of recreating our species. This also has become a way of finding increased momentary pleasure and has caused us to find many and varied interpretations of increased pleasure in this way over tens of thousands of years. Biologically, this has also become a cultural introspection, with rules and judgments prevalent.

In more recent history, tobacco was found in use by one family of our species and through transposition with each other has become another pleasure-enhanced state. Chemicals entered the drama much like tobacco. Momentary chemical enhancement was discovered by using heroin or cocaine or smoking certain plants or from mixing potions from the sea. Pharmacy has created another chemical derivative that causes a momentary, sensory enhancement. Of course, intermittently achieving success in our intended path of endeavor also gives this pleasure.

Evolution has molded us to maintain a neutral state of mind, with only bursts of sensory pleasure making us momentarily happy. Even if we were to constantly live a life in a toxic state of supposed pleasure, we would become immune to the lack of bursts of sensation and thus seek even higher and more thrusts into the sensory path. For most of us, evolution made it that these bursts of pleasure quickly subside, and we can maintain a banal disposition. However, these occasional bursts of dopamine have evolved us to create a better civilization. When we successfully succeed in completing a task and receive the reward in our brains, we are inclined to repeat this exercise for further suitable reward.

Aldous Huxley would tell you that happiness equals pleasure, and even some scientists today think that by spending billions on research of the brain we could produce constant happiness. The paradox of this is the anticipation of a positive outcome is what brings pleasure. The best example would be the love-hate task of raising of a child through diapers, education, and teen hormone changes to adulthood and happiness, for us who raised the child.

We are being told of the developing age of artificial intelligence, or AI. This would enable a computer -modeling scenario whereby computing technology theoretically produces a state of never-ending happiness. This would guide us into a euphoric state of semi-conscious underachievers looking for the best shopping article or holiday resort. This computing science would look after all our needs and predict our future with accuracy all without input from human touch.

Stop, Think, Discuss

Pleasure, as discussed herein, is also the realm of the new computer modeling industry. Called artificial intelligence (AI), this industry collects vast amounts of information from everything we *Homo sapiens* do every hour of every day. The hope is that AI will enable us to be happy all the time and perhaps direct our thinking, especially to products we think we desire.

Could AI be the next political system by its influence over us? Could AI take over our subjective thought process where theology has resided for thousands of years?

Question one: In these statements about creation, what is your opinion about the work of God and what is coincidence? If creation is an act of God, was it the way described in theological literature? Is this God's creation in ways other than literature or a random act of disorganized matter?

Question two: What is the *ontology* of the statements of happiness and the possibility of AI having great sway over us? Will we outperform the computer in the future, or will it control us?

Question three: Can you see the point of view of those who would not agree with your position on this subject? Does this ability to understand the other point of view matter?

Question four: If scientifically everything that happens has a precursor that makes it happen, why were *Homo sapiens* given the precursor of seeking happiness with chemical and other inducements.

Question five: On the scale of predatory animals, where do you think we sit?

Question six: Provide an opinion as to the correctness of this chapter. Use material at hand, if you have it.

CHAPTER 13

History Provides Simplistic Solutions for Chromosomes, Genes, and DNA

Ancient Greece

The story and evolving knowledge of genes, chromosomes, and DNA begins in ancient Greece, around 500 BC. The school of Pythagoras viewed the creation and subsequent development of a child as a variation of mathematics based on the triangle, which he is credited with for solving mathematical formulas. Outside of mathematics, he was mystical in his thinking.

His followers viewed the fetus as male-dominated via input of a sperm. The sperm collected mystical vapors as it traveled through the male body, which gave the infant its physical and mental attributes. Implanted in the female egg, it was nourished while growing into the fetus. Thus, the male and female role in the development of the baby was clearly divided. At birth, the infant was primarily the result of the sperm, which diminished the female role of nurture, clearly defining the male role of primacy in creating babies. The trilogy of human development was father providing, mother nurturing, and the resultant child. One hundred years later, Plato, a student of Socrates, wrote how a perfect child could be the result of exploiting the sperm of the father and nurturing of the mother into a perfect human being. This point of understanding, called spermism, permeated popular thought until the school of Aristotle intervened about three hundred years later.

Just this era of evolved and elevated thinking, which permeated everything in such a small part of the world, is worth taking a few lines to discuss at this juncture. This era occurred at a time when most of the rest of the planet was rubbing two sticks together to make fire. It was also happening in a very localized area of the world with a select group of people. This area, which we now call Greece, had culture and people in sufficient numbers to support this and other futuristic thinking along with variation of thought, which promotes quality of conclusion. They also had enough population to provide future progression and cultural stability. These great thinkers were an anomaly of the era for several hundreds of years. The only society comparable at the same time was the birth of the Buddha in Tibet and the Chinese culture progressing through societal change. In the context of the history of humanity, this is sudden, significant, and very quick in the progression of Sapiens. It is comparable to the Age of Cognition eighty thousand years ago.

As a philosopher and scientist, Aristotle pontificated on numerous subjects. His views of physics are now somewhat discarded as not very well founded compared to the school of Democritus. In history, he is not regarded as a great supporter of female recognition. Regardless, his observations of the development of infants and children have proven to be correct in history. He observed that children take on the attributes, both physically and mentally, of the mother as much as the father, and therefore he posited that the development of the fetus is the combination of both the sperm and egg. He used the analogy of a carpenter's hammer interacting with wood. He researched this with numerous cases and penned it into history as "material and message" through his teachings. For this, some would call him "the father of modern genetics", well before the diligent studies of Mendel and his peas in the mid-1800s. Philosophy and science were always human-centered and proven by physical observation by these ancient philosophers. This changed with Albert Einstein in the early 1900s.

The understanding of "message" was dispensed with much later in the dark ages once again. It became a belief that the sperm contained a mini-human, or *homunculus*, which needed nurturing to become a baby. The simplicity of the "new idea", called preformation, was that it recurred without a limit. At the beginning of the age of science, in 1520 an alchemist named Paracelsus believed that replicating a mother's womb with warm horse dung, placing sperm into it, and burying it into the ground to keep it warm would produce a baby, but with horrible characteristics. Perhaps this ideology led to the idea of Russian dolls that are packaged one inside the other. The idea of a homunculus provided a powerful doctrine for early Christians on the idea of original sin, since each of us were believed to be mini humans that had originally existed in Adam's body thousands of years prior. Thus, through direct replication, we were infused with the taste of the fruit of his original sin, eaten in contradiction to God's command.

This myth was so profound that it existed well into the 1700s. Even when microscopes were invented, scientists wrote of seeing small human beings floating in sperm. Sperm with heads and tails that formed other parts of the body were understood as commonplace. In philosophy, it is like the theory that you see what you want to see and hear what you want to hear. Theory had it that this small human just needed to be "inflated" to become an adult *Homo sapiens*. Not until 1768 did the idea of the sperm and egg containing a kind of code that provided instructions for making a human from scratch arise—again. This was the repetition of two thousand years of Pythagoras and Aristotle, discerning the theory of baby making. The whole discussion for two thousand years was implied as original thought by those ascertaining the meaning of life.

Christian scholars, biologists, and philosophers continued dialog about these two concepts right into the 1800s. For all the elevated dissertations, dogma, and reasoning, it was strictly a reprocessing of two concepts from the ancient philosophers Pythagoras and Aristotle two thousand years prior. Since that time, both theories have been both destroyed and vindicated. Both Pythagoras and Aristotle were partially right and partially wrong. However, in the 1800s, there was an impasse and scholarly minds were unable to progress beyond concepts of hundreds of years BC.

DNA is like a computer program but far, far more advanced than any software ever created.

Bill Gates, *The Road Ahead*

Stop, Think, Discuss

Just this slice of history in and around Greece, and in isolation, raises the question of how this philosophic and scientific thinking suddenly arrived on the doorstep of humanity so long ago. Could it be divine intervention in the evolution of our species?

In the process of evolution, one comment made is that we all come from apes and or monkeys. There are those who ask, "If we all came from monkeys, why are there still monkeys?" We humans have 23 pairs of chromosomes. Did you know that there are twenty-four pairs of chromosomes in a monkey or ape?

Question one: Do you think the Pythagorean idea, which was superseded by Aristotle, which in turn was superseded by a version of the Pythagorean model again in the Middle Ages, was due to clear thinking or for some other reason?

Question two: Group thinking is evident in the era of preformation. Can you see how even when thinking is predominantly believed by almost everyone, it can still be incorrect? Even after seeing evidence with their own eyes, with the invention of the microscope, of human creation by combining egg and sperm? Do you see theology giving a scientific explanation in this?

Question three: Where did Pythagoras get his idea from about the creation of an infant? How did his idea influence civilization, and was this a good influence or bad, looking back?

Question four: Is it possible that the idea of the creation of an infant in its mother's womb was a male-dominated process? Was leaving the mother's role to one of nurturing an idea that grew out of a male-dominated society?

Question five: Do you know how many elements make up a string of DNA in humans?

Question six: What other historical questions do you still have about the idea of creating an infant?

CHAPTER 14
Charles Darwin

Looking back, in 1830, the impasse between historical fiction and present-day fact, like genetic code and homunculi, began to unwind. Charles Darwin boarded the ship *HMS Beagle* and set his sights on biology. Originally studying for medicine, he left that to study theology in Cambridge. Although a very sincere theist, he continued to work in botany, geology, geometry, and physics. He was influenced by the philosophy of William Paley's 1802 *Natural Theology or Evidences of the Existence and Attributes of the Deity* and later John Herschel's 1830 *A Preliminary Discourse on the Study of Natural Philosophy*.

Paley compared the construction of a watch to the construction of *Homo sapiens*. If the gears of a watch were made by a master craftsman, so too are the ligaments of the hip joint connecting the leg to the body. This could only be the work of God, a master craftsman. Herschel reduced very complex phenomena to their cause and effect, something philosophers tend to do and which they call ontology. He was primarily interested in the creation of biological organisms, which he saw in two steps: the first being the creation of life from non-life and accepting the idea of divine creation; and, the second being that he wondered how, once life was created, it diversified over the ages. He posed the question "mystery of mysteries" in asking, how can the relics of the past determine the present?

Once again, we find the scholars of theology, biology, and philosophy at odds, and the discourse of natural history of living things devolved into the study of plants and animals without history. Darwin wanted to provide a study of natural history with the basic philosophy of causes and effects. Darwin asked how organisms transmit (or transmute) information from their origin over thousands of years. By the way,

Mendel, another Christian theist, asked how organisms transmit information to their offspring over one generation. Both questions look at cause and effect: Darwin over a long time and Mendel over only one generation.

While at sea, Darwin read Charles Lyell's 1832, *Principles of Geology*, which argued that geographic formations were created over vast periods of time by a slow mutation process like erosion, not by the hand of God. Lyell argued that God shaped the Earth not by a single cataclysmic event, but rather by millions of smaller events. This permeated Darwin's thinking—the idea that biological creation was an evolution of creation, evolving again and again, over time. Although Darwin gets credit for the theory of evolution, he was influenced by the thinking of others of his time.

Darwin proved to be a virtuoso in his ability to catalog and observe species. He discovered and cataloged dinosaur remnants, birds, animals, and plants. In the Galapagos Islands, he cataloged animals of the same species that had mutated to different coloring or different subtypes. They apparently ate the tortoises, and he was unable to determine if these too had been mutated from island to island. His genius was the ability to find patterns in biological order. So, when it was pointed out that several different-looking birds were one species that had mutated to several by the distance of island to island, he asked himself a question: What if the current animals were the consequence of millions of evolvement events over millions of years? Darwin dallied much too long to publish his findings. Another man, Alfred Russel Wallace (1823–1913) was busy completing nearly identical scientific findings with his theory of evolution through natural selection, independently of Charles Darwin. He sent Darwin a copy of his discoveries and jogged Darwin into completing his own work in time to present his catalog of findings to the scientific community jointly with Wallace.

With Alfred Russell Wallace at his heels, Darwin began his writings with a diagram of life from the roots, like a tree; not from a human, God-centered being with everything radiating from that origin. He used the knowledge of his youth from farming as a basis for his theory. Farmers used variants in their livestock to reproduce other livestock with new beneficial properties. By selective breeding, they produced better animals to suit their needs. Using his extensive reading, this time of another clergyman, Thomas Malthus, Darwin reasoned that selective breeding was also the result of the death of the weak and survival of the strong. The survivors breed the new stock and another breed is born for all time. About ten years later, another great mind was exploring these same issues, but prior to Darwin publishing his book. Alfred Wallace was nearing the same conclusions as Darwin and sent him his paper. Darwin presented both papers to the scientific community in 1858, so both would be credited for their hard work. Both papers were not particularly noted for any noteworthy discoveries.

Using new terminology, Darwin subsequently tried to explain the theory of heredity by combining the Pythagorean theorem and the Aristotle theory into one. He speculated that "gemmules" from the male were transmitted to the "germ cells". Germ cells are the sperm and egg and this phrase becomes much more important in later discoveries of the modern era. Darwin's gemmules carried instructions to build components of the human body. So, the Pythagorean little people circulating in the male body were carrying Aristotelian code to build babies in the womb.

Sadly, in his search for the truth, Darwin missed the writings of Gregor Mendel in 1856 while he was writing his new book on evolution. This would have immensely assisted Darwin's search for the truth.

Darwin was an observer. He cataloged his observations. He drew conclusions from observation. He was not someone who experimented.

Mendel experimented, documented, and did the math.

Stop, Think, Discuss

I was recently in the Galapagos and saw firsthand the giant land tortoises. They are amazing! They are about ten feet long slightly less wide and about five feet high. Can you imagine seeing these for the first time? And eating them on the voyage home? It must have taken half the trip to finish their turtle soup!

Question one: In these statements about Darwin's observation of evolution, is there any connection to the divine intention of God? Is God even involved? Is evolution God's creation?

Question two: What is the *ontology* of the statements? That is, what is the cause and effect, free from our personal bias, of these statements surrounding evolution? They contradict theological thinking at the time, but was this theology turning itself into science for interpretation of events?

Question three: What kind of personality would it take to contradict theological interpretation of science at the time of Darwin? Now that science is the accepted "theology," how would this same personality be treated today if contradictions of scientific theory were postulated?

Question four: Absolutely everything that happens has a precursor that makes it happen. Then the event happens. Darwin reading Lyell's principles of geology influenced his thinking greatly. Would he have come to the same conclusion had he not read this literature beforehand?

Question five: How would you describe Darwin, in your own words?

Question six: Are there issues in these statements with which you disagree? Explain.

CHAPTER 15

Gregor Mendel

Gregor Mendel (1822–1884) was a scientist, Augustinian friar, and abbot of St. Thomas' Abbey in Brno, Vienna. He studied all the sciences and loved botany. He wanted to be a full teacher in the Abbey school but was failed by his teachers and thereby ended up being a substitute teacher for the rest of his life.

Mendel tended his garden and planted another crop of peas. He had already bred them to be true, which means that they produced the same offspring with the same color, or the same seed texture. These purebred plants were the starting point of his experiments, revealing a pattern of discovery. His experiments were to produce hybrids. Tall plants bred with short plants—what would they produce? Intermediate, short, tall, or several of each?

Mendel spent eight years crossing hybrids with hybrids, carefully planning, planting, organizing, crafting, and caring for his peas. He discovered dominant and recessive traits. When he bred dominant traits with recessive traits, the recessive traits disappeared. Then, in the third iteration of the plant, the recessive trait reappeared, perfectly intact.

By using mathematics, Mendel was able to construct a model explaining the inheritance of traits. Every trait was determined by an individual particle of information. The particles came in two variants, one inherited from each parent. Only one particle would assert its existence. The other would lie dormant but intact. When these traits were crossed, the dominant particle would be the trait of the plant. However, when two plants were crossed that had recessive particles, the recessive trait would reappear as if it had never left.

He spent years tending to his crops. Crates of peas were shelled, crossed, and logged into his books. His discovery revealed the most important feature of a "gene", in that each gene was unique, separate, and fixed in its traits.

Mendel produced a paper on his discoveries. He was not a wordsmith, so his epistle made for heavy reading. It was presented to a small body in Austria and forwarded to the Royal Society in London, the Smithsonian in Washington, and numerous other influential bodies. He also forwarded copies to several scientists around the world. What followed was "one of the strangest silences in the history of biology," as stated by William Bateson (1861–1926), a fellow geneticist from England and popularizer of Mendel. From 1866, when it was published, to 1900, Mendel's work was lost to the world. Mendel died in 1884, having had to forego the work on his beloved plants for the work as the abbot of his school. When he died in obscurity, one of his students related, "Gentle, free-handed, and kindly ... Flowers he loved."

The greatest single achievement of nature to date was surely the invention of the molecule DNA.

Lewis Thomas[34]

But that is just the beginning of the story. Within a period of three months in early 1900, three independent botanists discovered Mendel's laborious work. Each had independently thought they alone were master discoverer of this new theorem and sought out previous works on the subject. Miraculously, each found Mendel's meticulous work in the library of science in Vienna. One of them, William Baetson, took only an hour to become a convert to Mendel. He made it his personal mission to ensure that Mendel would never again be forgotten. Baetson was a professor at Cambridge and knew he needed a name for this new science. He called it "genetics", the study of heredity and variation.

Baetson determined that "if genes were, indeed, independent particles of information, then it should be possible to select, purify, and manipulate these particles independently from one to another" (Mukherj 2016).[35] Scientists could possibly weed

---

[34] Encyclopedia Britannica https://www.britannica.com Lewis Thomas (born November 25, 1913 in Flushing, N.Y., U.S. and died December 3, 1993 in New York, N.Y.), American physician, researcher, author, and teacher best known for his essays, which contain lucid meditations and reflections on a wide range of topics in biology

[35] For extensive forward thinking and extension of thought in the gene biology, please read the book, *The Gene: An Intimate History* by Siddhartha Mukherjee, May 17, 2016

out undesirable qualities and change the "composition of individuals and nations" (Mukherj 2016). He further stated this profound thought of the future: "Whether the institution of such control will ultimately be good or bad for that nation or for humanity at large, is a separate question" (Mukherj 2016). With that elevated thinking we enter the period of eugenics.

Stop, Think, Discuss

Examine the pros and cons of new ideas. How do you think these ideas were being accepted by the majority of God-worshiping people?

How would you imagine the dialog of the negative side of this discovery would take place?

Question one: Mendel was a man of God and worked diligently at this task. Do you think God also had a hand in it?

Question two: What is the *ontology* of this study of genes? That is, what is the actual information, free from our personal bias, of these statements? Do you see this premise working for Mendel?

Question three: What appears to be the precursor in this? Was it an experiment created by God or by man?

Question four: Absolutely everything that happens has a precursor that makes it happen. Then the event happens. Why do you think this study came to the forefront in 1900, after being forgotten? Was there divine purpose behind the discovery? Or was it coincidence?

Question five: How would you describe Mendel, in your own words? Was this theological interpretation or pure science, at the time? Mendel experimented, and Darwin cataloged. Both have had significant impact on the world today. What do you think the significance of either or both is?

Question six: Are there issues in these statements with which you disagree?

CHAPTER 16

Theist Involvement in Evolution

Take a moment to think about theist involvement in this new understanding of the evolving of life. Islam had diverted from scientific endeavor in the fourteenth century to focus on the religious roots of their belief system, whereupon Christianity took over, led by Albertus Magnus (1200–1280), a German Catholic Dominican bishop, and his student, Saint Thomas Aquinas (1225–1274), another German Catholic Dominican bishop. Hindus, Buddhists, and others respectively had members of their faiths who were engaged in science, but not nearly as involved as the Christians. The Jewish faith continued in a tradition of not being evangelic, and therefore their close cousins, the Christians, inherited the role of power and rule—a role well earned after being absorbed by the Roman Empire in the prior centuries. Note, however, that Jewish scientists have been involved in some of the greatest scientific discoveries in history, and by a larger number of scientists than other cultures. Jewish emphasis on education and learning is the precursor for this impressive result.

The rules of order and law were mostly through potentates, and they, in turn, were controlled by Christian leaders' influence over them. The pope had ascended to the role of a God. While Martin Luther was creating a religion of the masses for the masses, his theological belief system was also centered on historical thoughts and doctrine, not in the new scientific world.

Suddenly, the world had new information, and it contradicted the learning of theologians well trained and versed in previous dogma. They were not going to give up their historical dogma without a fight, regardless of the scientific and observable truth. All manners of voice and prose were cast against this new thinking that was contrary to the historical teachings of God.

But not everyone complied. Clever minds dwelled upon the idea of selective creation. What would humanity be if they culled out the mistakes of birth? Those seemingly undesirable mutations with regressive human traits? Nature had managed a process through the ages to result in a stronger and healthier species. With human intervention, it was used to build better livestock and plants. Why not *Homo sapiens*, too?

Darwin's cousin, Francis Galton, pursued a plan for this very thing in 1883. He called it eugenics, and the name has continued to this day. Galton stated that: "Believing as I do, that human eugenics will become recognized as a study of the highest practical importance, no time should be lost in compiling personal and family histories" (Galton 1907). He immediately began looking at the human condition. Never mind the more acceptable scientific route of looking at the routine of plants. Why not start at the top of the list?

In relatively little time, Galton crossed swords with Bateson, the genetics professor, who was the champion of Mendel. After much fighting and discourse between botanists and geneticists, by 1909, it was determined that heredity was involved in all life, not just plants, and Mendel's laws were applicable to all of life. In 1909, the term "gene" was coined. The scientists who made up the phrase had no idea of its importance, just that there was a process whereby information from the male and the female produced an offspring using code of some kind.

Still, the idea of eugenics—the weeding and culling of the human garden—lingered. In 1912, a scientific gathering of the minds occurred in London. Two presentations stood out from all the rest. The first, a German presentation of "race hygiene" was a premonition of racially self-cleansing times to come. The second paper was from the United States of America. An energetic production, authored by Charles Davenport, a Harvard-trained zoologist, the paper eloquently spoke about the already planned efforts to eliminate defective strains within *Homo sapiens* in America. They even had government ministries in place for this very action plan.

Stop, Think, Discuss

Did you realize that it was America and Britain that were the world leaders in eugenics, but they quickly dropped it when Germany endorsed this before the Second World War?

The documented files of the residents of the isolation camps in America collate the sterilization of women (often prostitutes) and the harboring of political enemies of the state among other otherwise "non-normal" people. Can you see how the science of man, like some religious dogma, is quickly perverted to usurp control in a political hierarchy?

Question one: In these statements, determine what the divine intention of God is and what is the intention of man. Is God even involved? Is eugenics God's creation or man's?

Question two: What is the reason for the science of eugenics? Does this have scientific implications or theological?

Question three: With all the clarity of history, and attempting to understand a good side of this, how does one come up with the idea of eugenics with a positive view of it? Was this initially an evil idea or a misplaced ideal?

Question four: Absolutely everything that happens has a precursor that makes it happen. Then the event happens. Galton was the cousin of Darwin. Was he just trying to ride on the coattails of the genius of Darwin and could care less about the consequences of eugenics?

Question five: Looking backward, do you think the world would be so energetic about eugenics if they knew the consequences beforehand? If you were not a prophet, could you predict the result of this scientific undertaking? If you opposed eugenics at that time, who do you think your opponents would be, and how committed would they be to their positive resolve of eugenics? Is this an example of groupthink or something else? Explain.

Question six: How would you or your relatives react to this science if you were alive at the time?

CHAPTER 17

Inferior Genes

The United States would create confinement centers for the genetically unfit. Committees were in place to consider sterilization of unfit men and women, epileptics, criminals, deaf mutes, the feebleminded, those with eye defects, dwarfism, bone defects, schizophrenia, manic depression, or insanity. Nearly ten percent of the population was inferior, according to the US. There were already laws authorizing sterilization in eight states. That they had found no ill effects of sterilization was stated by California in 1912.

In 1924, the state of Virginia had colonies for feeblemindedness. This classification consisted of three categories—idiot, moron, and imbecile—as documented in files of the various states of America.[36] An idiot was easiest to quantify, as it represented a mental age of 35 months. Imbecile and morons were more subjective and were supposed to represent mental cognitive disability. In practice, it was a category that allowed severe interpretation of individuals who were at odds with society, often with no mental illness at all, including prostitutes, orphans, depressives, vagrants, small-time criminals, schizophrenics, dyslexics, feminists, incorrigible adolescents, or anyone whose behavior, appearance, or choices were abhorrent to the norm. Feebleminded women were sent to confinement to ensure no further breeding, thereby contaminating the population with morons or idiots. Many of these women were sterilized, but only after court authorization. This mindset had history in American slavery with the historic fear that African slaves, with their inferior genes,

---

[36] Idiots—Those so defective that the mental development never exceeds that of a normal child of about two years. Imbeciles—Those whose development is higher than that of an idiot, but whose intelligence does not exceed that of a normal child of about seven years. Morons—Those whose mental development is above that of an imbecile, but does not exceed that of a normal child of about twelve years. — Edmund Burke Huey, Backward and Feeble-Minded Children, 1912

would intermarry with whites and thus contaminate the gene pool. In the 1860s, laws were introduced to prevent what was perceived at the time to be a travesty and served to calm the population. However, immigration magnified these anxieties again in the 1920s, but this time, not with African slaves, but with European immigrants. There became a feeling that immigrants would flood America and cause "race suicide". Many other Western countries agreed. Baetson, the geneticist and champion of Mendel whom we discussed in the previous chapter, wrote that "eugenic ravens croaked for reform in England. America, by comparison astutely howled like wolves." They were cleansing humans by culling out the mutations. Counterbalancing this was the idea of genetic purity.

Sixty-two years. That is all it took for *Homo sapiens* to go from the discovery of Mendel and his peas, to the concept of sterilization and purification of humans. As the American eugenics' movement advanced, almost all of Europe watched eagerly with envy. By 1936, a virulent and macabre form of eugenics was overtaking Europe.

Five months after gaining power in Germany, Hitler passed the sterilization law. Borrowed from the Americans and greatly amplified, this law enabled the sterilization of those with hereditary diseases, including mental deficiency, schizophrenia, epilepsy, depression, blindness, deafness, and serious deformities. This law allowed that, once approved by a court, the person was to be sterilized even against their will and force could be used if necessary.

All while this was happening, the counterbalance theory was the perfect Aryan body. Then, in 1933, Hitler also brought in a law that allowed sterilization of dangerous criminals, which included political dissidents, writers, and journalists, by force.

By 1935, the laws also included Jews, who were forbidden from intermarrying Aryan people, thus contaminating their race. By 1934, nearly five thousand adults were sterilized per month. Finally, in 1939, Hitler moved to outright euthanasia of people with "defective genes". This was applauded by certain political and scientific elements in America. Thus, began the extermination of the Jews in Germany over the next several years, and as originally detailed in Hitler's 1925 book *Mein Kampf*. Try to place yourself in the mindset of Germans at that time and ask how and why it is that the Jews became so disenfranchised in the "mind state" of that time or even numerous times previously in history with other nations?

While Germany was conducting euthanasia, Soviet Russia saw the eugenics program in an entirely different perspective. Their scientists felt that genes would be active in the brain and could be brainwashed out of their system. They used electric shock therapy on their subjects. Any scientist who disagreed with the existing establishment was put to the Gulag and brainwashed until they complied with their masters.

In Canada, the population was enamored with the idea of assimilating the Native population into the "Canadian" population. In 1840, under British rule, they commenced the creation of residential schools, "government sponsored religious schools established to assimilate Indigenous children into Euro-Canadian and Christian culture" (Miller 2018). These residential schools morphed into popular thinking, like the rest of the world in the field of eugenics.

Generally accepted is that residential schools were established after 1880, but the first was opened in 1840 in Manitowaning, Ontario, under the colonial reign of Queen Victoria. The last residential was closed as late as 1996.[37] (Miller 2012)

The government officials took Native children from their homes and communities and placed them into central boarding schools, where they were forbidden even to speak their native language, including during the private moments of their residential tenure. Punishment was meted out in physical form for breaking these rules. Lives were destroyed, families broken, and grieving parents longed for their children who were prisoners in custody. Children's spirits were broken, family concept and child raising disappeared, and the result is that today, potentially five or more generations later, depending upon where they lived, there is still latent tragedy from the alcoholism, incest, drugs, and mental breakdown of the survivors and their offspring in several instances. Many of those who experienced the residential schools lost their ability to parent children in their own ways and that is perhaps why there is a much higher incarceration rate among the Native population than there is in the rest of the population.

---

[37] The Canadian Encyclopedia, Residential Schools in Canada, by A.J. Miller https://www.thecanadianencyclopedia.ca/en/article/residential-schools

Stop, Think, Discuss

Madness evolved from the great scientific work of Darwin, Walker, and Mendel. Science was taken by the minds of controlling totalitarian regimes and used to gain power over others. Junk science is also the foundation of most totalitarian regimes, and in turn, they produce junk science, further enhancing their power.

Can you think of any other scientific or religious ideas that were usurped to prop up power?

Question one: Prior to this reading, did you know of the involvement of the United States and Great Britain in the role of culling our species?

Question two: Can you think of any rational thought process that would allow numerous scientific minds to think of culling our species? What was the rest of the population aware of and what were they thinking?

Question three: Could there be an element of God in all this thinking? Was it an idea created by God or by man? Could a small number of proponents in opposition have changed any of this thinking or had groupthink taken ultimate control?

Question four: Absolutely everything that happens has a precursor that makes it happen. Then the event happens. What happened, either politically, theologically, or scientifically, to mandate the process of eugenics in the world? With the philosophy of 20/20 hindsight, would you blame science, politics, theology, or something else for this horrific course of events?

Question five: Describe this chapter in your own words.

Question six: Are there issues in this chapter with which you disagree? Explain.

CHAPTER 18

The Discovery of the Gene

The gene was a mathematical statement, but something unseen in practice. Beatson traveled to the United States in 1907 and met Thomas Morgan, a cell biologist who worked as a professor of zoology at Columbia University. Morgan wanted to uncover the physical basis of heredity. He wanted to see the evidence through cell multiplication and genes at work. It had been previously proposed that genes lived in chromosomes in the nucleus of the cell. It had already been seen that the Y chromosome was only present in males. Also identified was that genes were residents in chromosomes, but the exact science was unclear.

Like Mendel, Morgan started to experiment, but this time with fruit flies. He discovered an important modification to Mendel's laws. Genes do not travel separately but rather in packs. Packets of genes were packaged into chromosomes, and ultimately, into cells. With this, Morgan linked cell biology and genetics. The linking of genes within chromosomes and the unpinning of them also eventually led to our present understanding of DNA. The fruit fly superseded the pea as the fertile soil of discovery

Every cell and every organism needs information to carry out its function. Where does the information come from? Genes carry that information. The developing message carried by the embryo of an embryonic child is the gene. It is the instructions that allow an embryo to build an ear or a hand. The gene is your inheritance. By the 1940s, this pair of molecules connected by hydrogen bonds had still not been discovered.

It was understood by then that chromatin carried protein, and protein is very important to the body. For instance, within the body, sugar combines with oxygen to make energy and gives a byproduct of carbon dioxide. This chemical reaction needs a stabilizer to keep it from consuming our bodies, and proteins are the source of the stabilizer. Also, a new chemical became understood as an acidic component of our cells. It was named nuclein, which later was renamed nucleic acid, and it comes in two forms: deoxyribonucleic acid (DNA) and ribonucleic acid (RNA). Both are long chains made from four components called bases, which are strung together like a backbone.

Oswald Avery wrote a paper in 1944, describing his experiments of discovery with our DNA. He proved that our heritable traits are carried by our DNA and RNA, and that eugenics was a sham. This was at the same time as the apex of the extermination camps in Germany. Sadly, eugenics as a political movement continued until the early 2000s with sterilizations that were, until that era, usually court approved.

Stop, Think, Discuss

Combining genes and that chemical process within our bodies causes me to reflect upon the combining of sugar, protein, and oxygen to make energy in the human body with a byproduct of carbon dioxide. Scientists indicate that each human produces nearly one kilogram of carbon dioxide per day, unless engaged in a physical effort, whereupon it can go to about four kilograms. Assuming everyone leaves their bed, let's assume two or more kilograms per day.

Now let's jump into only slightly connected thoughts in this chapter; thoughts that originate with the gene, oxygen, and sugar, and continue well past that explanation, thereby admitting to a spurious postmodern thought process that might seem unconnected yet significant. Let's call this one an author's jump-start!

Think about global warming, as it is discussed regularly these days. One generator of carbon dioxide is the human body, via the reaction described above. In the year 2019, with approximately 7.5 billion humans on this orb, the human body produces significant carbon dioxide. By calculation noted from figures above, it would be nearly 15 billion kilograms of carbon dioxide per day. Work that out to a year and you get 5.3 trillion kilograms of carbon dioxide in the atmosphere from humans just being alive and healthy. Massage the numbers up or down with your opinion about how much we exercise through work or play or just being alive and out of bed.

Now, compare the human environmental input of carbon dioxide to other reactors normally accosted with this environmental issue. Every petroleum-burning passenger vehicle produces about 4,600 kilograms of carbon dioxide per year. Popular opinion states that there are over one billion passenger vehicles in the world. So, the math would indicate humans produce more carbon than the family car.

Then an argument comes that humans eat carbon-producing material and thus do not impact the atmosphere in any significant amount. But don't petroleum products originate from these same fundamental sources also?

Carbon dioxide can be measured in the atmosphere. That is factual, as possibly is the effect of carbon dioxide upon the life of the planet, although the information varies as to the environmental group that produces the numbers. Think about the statement by the United Nations that the ice fields in the Himalayas were melting at a phenomenal rate, a statement that was corrected years later because of the exuberance of those producing the information.

Do you think that humans are causing some climate change just by being on the planet and living and breathing? If you say "yes", how do you think you would deal with this without genocide, or eugenics, or some other idiotic idea?

Do you require faith to believe this science and, if yes, is it crossing over into the field of a new theology?

Within climate change theory, has absolutely everything that could affect climate been inserted into the computer model? Can every future item affecting this computer model be predicted in absolute terms with accuracy? If yes, then explain your thoughts about meteors, volcanoes, and solar events like the eleven-year cycle of total solar irradiation.

Question one: Mendel's work was initially modified in 1907 by Thomas Morgan. Do you see this as natural progression or an act of God? Is this God's process or man's?

Question two: Do you see scientific work here free from personal bias?

Question three: What appears to come first in this? Was it a theological process or purely scientific discoveries by man?

Question four: Absolutely everything that happens has a precursor that makes it happen. Then the event happens. What is the precursor or world event for this process to evolve?

Question five: How would you describe this chapter in your own words?

Question six: Has Harry Potter— creative thinking—taken over any of the dialog on this issue?

CHAPTER 19

Biophysics and Cell Biology

Biophysics, applying the laws of physics to biological phenomena, was born in 1946. Sticking with cell biology in this discussion, this science started in 1919 with Ernest Rutherford, a New Zealand physicist who took keen interest in this new science of DNA. He reasoned that the molecule was made up of more particles, which transmitted heredity. He was followed by Maurice Wilkins, another physicist from Cambridge, who graduated in 1930, and who headed up the investigation in 1946, in King's College, London, England.

By 1953, numerous scientists were working toward solving the structure of DNA. Mostly the United States and England were the homelands of these learned scientists. Two of them, James Watson and Francis Crick, in Cambridge, England, were able to make a helical model from molecule pairs. They discovered that the A and T pair opposed the G and C pair to create the right-handed, upward-twisting helix of two matched pairs of strings of DNA, laced together in a ladder of the molecules.

Wait, your eyes are glazing over. We are going to try to make this interesting and readable for those of you not in the industry. If you are in medicine, this is way too simplistic, sorry.

The size of this one molecule of DNA is one-thousandth of one-thousandth of a millimeter. One million of them stacked together would fill this letter "O". It is immensely long and thin. In every cell in your body, you have two meters of it. Drawing it as a scaled-up model as thick as a thread would result in a drawing that is 200 kilometers long.

Each strand of the helix is a complement of the other, locked with the A to T and the C to G, like a spiral staircase of information. It is the most important molecule in biology.

On April 25, 1953, James Watson and Francis Crick published their paper, "Molecular Structure of Nucleic Acids", which described a structure for DNA.

> Every time you understand something, religion becomes less likely. Only with the discovery of the double helix and the ensuing genetic revolution have we had grounds for thinking that the powers held traditionally to be the exclusive property of the gods might one day be ours. James D. Watson (Quotes 2007)

Once again, the world was given scientific knowledge heretofore unknown in the design and creation of life. The gift of knowledge is an understanding, which is conclusively proved by extensive testing and irrefutable. It is a gift given to *Homo sapiens* through the investigation and mastery of seriously intelligent people. There are those in our society that would call this a coincidence, while others would say it is a revelation, all of which is dependent upon your own perspective. Then we also must wonder about our idea of equality. How is it that these seriously intelligent people can think in the abstract of this issue and come to results that are oblivious to the rest of us? Where is the equality in that?

How is this gift of DNA structure to be perceived? Is it a divine gift, a coincidence of time, or just the unrelenting persistence of man? Who will use this gift and how? Look at what the discoveries of Darwin, Mendel, and Watson led to with the use of eugenics to cull those who did not meet the norms of society.[38]Let's enter the world of the medical student, the bioscience world of medicine genetics and so on. I am just going to give a brief example of the language and science they use in everyday discussion among themselves. We, the outsiders, are most likely going to start looking at our shoes a lot and yawn, because it is a world of science that has evolved almost into a world of understanding only within itself. The next few paragraphs are a peek into that world. This will not be a deep, long explanation of how this science works other than to give a glimpse of the inner workings of those scientists.

---

[38] For a great unpretentious book, an easy read without the author taking credit for everything, read, *A Crack in Creation* by Jennifer Doudna and Samuel Sternberg, 2017. Publisher: Houghton Mifflin Harcourt. *"Not since the atomic bomb has a technology so alarmed its inventors that they warned the world about its use. Not, that is, until the spring of 2015, when biologist Jennifer Doudna..."* Go get it and enjoy.

By 1958, the Nobel Prize in Physiology or Medicine was awarded jointly to George Beadle (California Institute of Technology) and Edward Tatum (Rockefeller Institute for Medical Research, New York) and Joshua Lederberg (studied under Tatum and taught at University of Wisconsin) for investigation on how the gene transforms through protein into function. The gene encodes information to build a protein, and the protein actualizes the form and function of the organism. Every mutant created is missing a single metabolic function, corresponding to the activity of a single protein enzyme. And further, every mutant is defective in only one gene, better known as the one gene–one enzyme hypothesis.

How did the gene encode information to build a protein? The protein is built from twenty amino acids, strung together in a chain. Unlike DNA, protein twists and turns irregularly in space. Because it is shape shifting, it allows the protein to execute diverse functions in cells. They become globular in shape to enable chemical reactions or become long and stringy fibers as in muscle. They carry information to build color in eyes, or flowers, or become a transportation vehicle for other elements such as hemoglobin, and so on.

Think of the coding strand of DNA or its messenger RNA as instructions for building protein chains out of amino acids. There are twenty amino acids used in making proteins composed of amino acids. They are represented by letter codes A, C, T, G, U and so on. Without going into a lecture of DNA and with an intent to give a cursory overview, just know that the use of letters to identify each amino acid is the norm. For instance, T is Threonine, H is Histidine, C is Cysteine, and so on. But how do the sequences of DNA and RNA—ACT and G in DNA, and ACU and G in RNA—carry the instructions to build the protein?

By 1960, Jacob, Brenner, and Crick, at an informal meeting, suddenly "rediscovered" the unique RNA first found in 1956 by Elliot Volkin and Lazarus Astrachan. They discovered that the DNA needed to be converted into another copy of itself, and it was this copy that translated into a protein. This copy was the RNA, the DNA's molecular relative.

Through transcription, the original DNA gives information to as many RNA molecules as necessary for the process at hand, and the RNA in turn gives information to the protein for function. This allows for as many copies of a gene as necessary to be in circulation at the same time, and for the RNA to be increased and decreased on demand. With this, a critical component of a gene's activity and function was identified.

But how does the transcription of ACT to ACU work? (ACT being the gene and ACU being the RNA, which is the copy in chemical terms.) It was then determined that it was the precise sequence of amino acid codes that determined genetic codes. This knowledge became vital to future investigations and scientific meddling yet to come. Like the English language use of ACT to be TAC or CAT, the sequence of amino acids is critical in heredity:

> The secret of DNA's success is that it carries information like that of a computer program, but far more advanced. Since experience shows that intelligence is the only presently acting cause of information, we can infer that intelligence is the best explanation for the information in DNA. (Wells 2006)

Stop, Think, Discuss

By 1961, numerous laboratories had become involved in a race to decipher the human genetic code. By then, it had been determined that the future of medicine and human development was clearly linked to DNA and that understanding the codes would potentially solve human disease.

It would also make some very wealthy. Money is a prime motivator and the rush was on to decode human DNA.

How do you think money and power motivate, and is this a good or a bad thing?

Question one: If God is omnipresent, does God have a hand in these discoveries? Is God even involved? Is this God's creation or man's?

Question two: How do scientists imagine molecules that are so small that if they were drawn to the thickness of a thread, they would be 200 kilometers long?

Question three: What are the odds of all the billions of permutations and combinations of DNA being a coincidence?

Question four: Medicine has historically been the use of chemicals to resist or prevent disease. There are about 250 chemicals that react safely with the human body to produce healthy results. Think of all the millions of medicines that are possible if protein were the chemical cursor that was used as medicine instead?

Question five: What do you think the future is for DNA medicine?

Question six: Do you see faults in the thinking of this branch of science? If yes, what faults, and what are the ramifications?

CHAPTER 20

Mutation

So how does like beget like and how do mutations occur? For mutations to occur, organisms need to generate variation, which results in descendants that are radically different from the parent. If genes transmit likeness, how can they transmit unlikeness?

One way of generating variation is mutation, or alteration, in the sequence of DNA, which changes the structure of protein and thereby alters its function. This occurs due to chemicals, X-rays, or when the DNA replication enzyme makes a mistake in copying the genes. Also, genetic information can be swapped between chromosomes. DNA from the paternal chromosome can swap places with DNA from the maternal chromosome at the time of generation of the sperm and egg. This can also recombine in future generations, resulting in a backward mutation.

This movement of genetic information from one chromosome to another only occurs in extremely special circumstances. It is when sperm and egg are generated for reproduction, and just then before the egg or sperm generate, the cells turn into a playpen for genes, as described by one scientist. The paired material and paternal chromosomes bind and readily swap genetic information, which is vital in the mixing and matching of heredity information between parents.

Damaged DNA can be replicated by its twin in the helix, resulting in the creation of hybrid genes. The right fixes the wrong. The original DNA is fixed, stitched back into the helix, and all is good.

In the 1950s, the discovery of "effector" genes showed that certain genes caused a sequence of development to occur only in time for certain phases of embryo development. For instance, the development of the eye is not necessary until there is a head, or a wing is developed only after the attachment points on the body are there and waiting. The effector genes kick into action only at specified times and specified locations and determine the identities of segments and organs. These master regulators work by turning other builder genes on and off. If the master is deformed, it could trigger a signal to build a set of toes onto a shoulder, etc.

What programs the master gene? What process tells it to sequence the structure of the head before the leg in the embryo stage?

By 1990, it was determined that this was a two-stage process. First, the mother produces an egg that has internal forces that cause it to accept the request from the mother's womb that the appropriate cells of the embryo migrate to the appropriate places so that organs and appendages and the like all end up where they're supposed to be, head up, toes down. These chemicals are mostly protein, deposited in concentrations relative to orientation, like loading the dice in a fixed gambling game. The chemicals are placed in a gradient relative to the head or toes. Head chemicals to the top, toes to the bottom, and the rest where they belong in between, like a map.

The discovery of the second stage was intensive, and in the process of discovery, in 1985, genes were found that cause cell death. Programmed death that is not by disease or fate or accident, but by plan. Then, it was also discovered that this programmed cell death was responsible for the cells that cause cancer. Most of these cancer cells were from ancient times, inherited by us for millennia. For example, activation of the gene BCL2 results in a cell in which the cell death cascade is suppressed; while this process is necessary for proper cell functioning, mutation may create a cell that is pathologically unable to die: cancer.

So, it is the relationship between the cells that causes protein to build in vast amounts or less, according to the connection of one gene to the other, and to which end the cell is oriented. It is quite simple, really, but when vast numbers of genes relate to each other, the equation creates vast variances of cell construction. To use an Aristotelian analogy, think of a carpenter building a house and another building a boat. Both use the same materials. The relationship of one board to the other determines the outcome. Human physiology is the developmental consequence of certain genes intersecting with others in the correct sequence, in the correct place. A fascinating read is the book *The Gene* by Siddhartha Mukherjee M.D., which once started, you will be unable to put it down until finished.[39]

> *A gene is one line in a recipe that specifies an organism. The human genome is the recipe that specifies a human.*

Richard Dawkins[40]

By the 1970s, biologists had begun to decipher the mechanism by which genes were used to generate organisms. They also determined that deliberate change in organisms was conceivable, and soon.

Viruses were studied for genetic code, and some were found that could enter the genetic code of other organisms and cut and paste themselves into the gene of the new organism and recreate themselves over and over. How did they cut into the mother's gene? Research determined that they did so with enzymes that cut and pasted the new gene into the DNA of the organism. It manipulated the gene to create a new gene.

With this knowledge, scientists began the work of manipulating the DNA of organisms and started building new DNA by cutting and pasting DNA from other genes into it.

Because the thing about viruses is that they're easily manipulated. The DNA they inject doesn't have to be destructive. It can be replaced with almost any kind of DNA you want, and it can be programmed to only replace certain parts of the host's genetic code. In other words, viruses are perfect vectors for genetic engineering.

> *Volumes of history written in the ancient alphabet of G and C, A and T.*

Sy Montgomery [41]

---

[39] Siddhartha Mukherjee is an Indian-American physician, biologist, oncologist, and author. He is best known for his 2010 book, *The Emperor of All Maladies: A Biography of Cancer* that won notable literary prizes including the 2011 Pulitzer Prize for General Non-Fiction, and Guardian First Book Award, among others. His 2016 book *The Gene: An Intimate History*, was among *The New York Times'* 100 best books of 2016, and a finalist for the Welcome Trust Prize and the Royal Society Prize for Science Books.

[40] Clinton Richard Dawkins, FRS FRSL, is an English ethologist, evolutionary biologist, and author. He is an emeritus fellow of New College, Oxford, and was the University of Oxford's Professor for Public Understanding of Science from 1995 until 2008. He is also an avowed Atheist and writes extensively upon that subject.

[41] https://www.goodreads.com, 82 Quotes by Sy Montgomery: Search for the Golden Moon Bear: Science and Adventure in Pursuit of a New Species

A great example of scientific progress is insulin. Previously, insulin was made by throwing animal organs into a barrel to ferment and drawing off the chemical discovered by Banting to counteract sugar issues in diabetes. Suddenly, the DNA of insulin was decoded, and insulin was created from a chemical process that does not involve animal organs. Insulin became safe, sterile, and pure, manufactured in any quantity desired. The inventors of this process applied for and got a patent for the process. Genentech shareholders have become incredibly wealthy since that day in 1978.

Suddenly and significantly, genetic engineering has exploded through the gateway of medical history into the current thinking of man. Knowing what could be done, it now became a sequence of how it could be done, and imperative to that was knowing the sequence of the human genome.

The initial goal was altruistic. Possibilities of its application included preventing disease, reversing disease, finding a cure for cancer, developing disease-resistant plants, creating a food chain that was reliable, and enabling the future of *Homo sapiens* to flourish. At least, this is my opinion, but when looking at the history of man, this notion is often bypassed for greed, power, and personal gain.

Through a long process of discovery, enhanced by financial gain to those who made the discoveries, the human genome was resolved. Great wealth was inherited by those who discovered medical treatments resulting in the sale of medicines. This encouraged others to join in the search for yet more discoveries. This is the infant state of the DNA research of today.

The future holds wonderful opportunities. Can cancer be cured? Can *Homo sapiens* live much longer and healthier lives? Can genetic diseases be rectified at birth?

Since humans have a propensity to make everything central to human activity, and to hijack knowledge to the benefit of one human group or another, what is the risk of this knowledge? Just like Darwin, Walker, and Mendel with their discovery of the human heredity and evolution, how will this be used to pervert human endeavors?

The University of California, Berkeley, researcher Jennifer Doudna, and Emmanuelle Marie Charpentier created CRISPR. CRISPR is the amazingly simple mechanism with which DNA can be cut, pasted, and modified.

Jennifer Doudna's 2016 book, *A Crack in Creation*, expressed her fears of the use of this knowledge. It is a book well worth reading and which could portend the future of this science, both good and bad.

Very soon after Doudna's paper on the technique appeared in 2012, labs all over the world tried it and found it was surprisingly easy to use. When something like this happens, it is difficult to sort good wishes from fact, but the prospect of CRISPR being an authentic process is intense. Doudna does, though, sound many notes of caution. Anyone can set up a CRISPR lab for approximately $2,000. Each CRISPR sample used to edit DNA will cost the amateur biologist $65 to have conveyed directly to their door. This opens the possibility of anything in human or biological development being tampered with, and there is no regulation of the endeavor by authorities. Think eugenics, Hitler, and the purification of the Aryan race. Then add to your imagination the potential for enormous amounts of money to be made by this process.

Currently, CRISPR is the most accurate form of gene editing. It isn't perfect. Three billion bases within the human genome portend the possibility of an improper base match of the twenty-base sequence necessary to cut and replace the imperfect sequence that would result in a permanent change in the human genome. Should gene edits only be allowed outside the body, as a safety precaution, or should we allow the edits of sperm and egg? Allowing the latter would change the germline forever. Once changed via sperm and egg, there is no going back. Doudna is cautious but has indicated her thinking that it could be in the sperm and egg, like Great Britain, noting that mitochondrial replacement therapy, which also leads to permanent germline alteration, is already a reality there.

Germline DNA is the source of DNA for all other cells in the body. The body's natural information is transmitted to germ cells—sperm and egg—which is then passed on to the new offspring.

Reading from newspaper articles, currently the most exciting potential medical application is in single gene diseases, like cystic fibrosis, sickle-cell anemia, and muscular dystrophy. This is the simplest possible task for CRISPR. Just one base must be corrected out of the three billion, and it's not hard to find. CRISPR can repair it. There is hardly an area of medicine that could not benefit from CRISPR.

There is also the dinosaur fantasy, kept alive by George Church at Harvard[42]. He is editing the elephant genome to create a creature like a woolly mammoth. Or

---

[42] George McDonald Church is an American geneticist, molecular engineer, and chemist. As of 2015, he is Robert Winthrop Professor of Genetics at Harvard Medical School and Professor of Health Sciences and Technology at

perhaps we could model Marvel comics and create beings that can breathe under water through gills, or chickens that weigh 20 kilograms. Perhaps a salmon that achieves five times natural body size in half its normal life span? Oh, yeah, that one has already been done![43] The Massachusetts company AquaBounty Technologies created this fish and is expecting FDA approval, because scientific review found almost no concern about risk to human health or to the environment. AquaBounty postulates it has done everything possible to show that farmed GM Atlantic salmon will be safe for consumption and for the environment. Many consumer groups disagree and are protesting against it and attempting to testify that it is poison on our plate. They want it labeled much like cigarettes are characterized. Perhaps this is a great misunderstanding of the science of genetic modification, but it points to the fact that this science is so far ahead our understanding that some of us are afraid of it.Great news recently came out on January 3, 2018. While driving and listening to the radio, I heard that: "A young man with an impairment in a gene that causes blindness called RPE65 was given a repaired DNA injection. Twenty-four hours later he can see perfectly." Can we believe this news? It does portend the ability to cure disease quickly and easily. Perhaps it is too late for some of us in later life, but for younger people, it is a possibility of curing disease without scalpel and pain of healing. But there is always a dark side, and it is unknown currently.

Harvard and MIT. He was also a founding member of the Wyss Institute for Biologically Inspired Engineering at Harvard.

[43] In 1992 scientists Choy Hew and Garth Fletcher figured out a way to add a growth hormone into Atlantic farmed salmon. It is created from Chinook salmon (*Oncorhynchus tshawytscha*), plus a fragment of DNA from the ocean pout (*Zoarces americanus*), an eel-like creature that inhabits the chilly depths off the coast of New England and Eastern Canada. This genetic code acts like an "on" switch to activate the growth hormone. The result was a genetically engineered super fish that grew nearly twice as fast, on less food, as conventional salmon. https://www.greenbiz.com

Stop, Think, Discuss

It becomes possible to create new mutant species in twenty-four hours, unlike Darwin's birds that took thousands of years to mutate. When the theologians, philosophers, and scientists discuss this, how will they try to control or use it from their own perspective or culture bias?

Who do you think will benefit from DNA science, both good and bad. Explain how?

The change in DNA of some food crops by Monsanto has already produced a worldwide negative reaction with protests and reviews. Do you think altered DNA is going against God's intention or producing poison within our food chain?

Question one: In the science of DNA is there intention of God leading man or is this only the work of man without any influence of spiritual intervention?

Question two: What is the information in this study of DNA that is free from our personal bias, and what is the knowledge, as learned by us, including our preconceived notions of these statements? Give your opinion about DNA altering the food chain or the human genome, medically and for gain or profit.

Question three: What appears to be the cause for the study of DNA? Was it an idea created from the study of eugenics or something else?

Question four: Could this study lead to aberrations in the human genome? Could *Homo sapiens* be changed forever by experimenting? Is there a medical reason to continue with the science of DNA? If this answer is "yes", how long do you think man should be able to live on this planet?

Question five: Describe your opinion about DNA food science and *Homo sapiens* science to someone else.

Question six: In your opinion, what is in these statements with which you disagree?

CHAPTER 21
Managing the Future of DNA

Control

Asilomar II, one of the most unusual meetings in the history of science, was organized by five scientists in February 1975. Not only scientists were asked to join the conference, but also lawyers, journalists, and writers. Looking back, it was an organizational stroke of genius. Opinions were to be gathered from scientists, as well as brilliant minds in other fields. Remember, this was at a time when Watergate was on everyone's minds and the perceived suspicious activity of government supporting and covering up immoral and suspicious activity.

Fierce debates on gene cloning erupted. The discussion surrounded the ridiculously simple method of producing genes that could ultimately alter the human species. A discussion about having a moratorium on experiments in this technology was the goal, recognizing the risks of this technology. Should scientists be restricted in their experiments on recombinant DNA?

The resulting discussion subsequently brought the lawyers into the fray. They issued forth about how the recumbent DNA going awry would bring forth draconian regulations, far worse than what the scientists would impose on themselves. This was the turning point of the meeting, and on the last day, it became apparent that the meeting could not end without formal recommendations, most likely because of the recent Watergate mess.

A four-level scheme of ranking the biohazard potential of genetically altered organisms was created: minimal, low, moderate, and high risk. Quoted from the Belgian Biosafety Server;

+ The required containment should be an essential consideration in the experimental design;
+ The effectiveness of the containment should match the estimated risk as closely as possible.

> Scientists established a classification of experiments involving recombinant DNA in order of increasing risk to human health and the environment. Four risk levels were identified: minimal, low, moderate, and high risk. A series of increasingly drastic measures corresponded to these risk levels, designed to limit as far as possible the release of recombinant DNA organisms into the environment. Good laboratory practices as well as the training of workers comprised the basic measures for any handling of recombinant DNA. The necessary physical containment measures were also described.

Second Asilomar conference 1975[44]

For any experiment, regardless of the level of risk, it was recommended to use biological containment barriers by choosing, for example, host cells and vectors that could not survive in normal environmental conditions. Certain experiments were simply forbidden: cloning of DNA derived from highly pathogenic micro-organisms or coding for toxins and large-scale experiments using recombinant DNA coding for products potentially harmful to humans, animals, or plants.

Recommendations generated a method of containment of each of the four levels of a hazard. First and foremost, agreement was that inserting a cancer-causing gene into the human virus would cause the highest level of containment. The other levels were also formulated with the ability to loosen or tighten regulations, as required in the future.

A minimum risk level of containment would be experiments in which the biohazards could be accurately assessed and were expected to be insignificant. Low-risk containment would be required for experiments generating identical genetic types, where recombinant DNA could not alter behavior of the recipient or increase its ability or even prevent treatments of any resulting infections. A moderate risk level of containment was for experiments where there was a probability of generating an

---

[44] https://www.biosafety.be, Biosafety worldwide—Historical background, Recombinant DNA technique

agent with the potential for ecological disruption. As stated, high-risk containment was intended for experiments in which the potential to harm the host of the modified organism could be severe and thereby pose a serious biohazard to laboratory personnel or to the public.

The proposal was almost unanimously accepted. The conference has since become known and is referred to as the Asilomar Conference. It soon became understood that "scientists were capable of self-governance". Those who were unused to this restriction were now required to control themselves.

This is the scientific reasoning on the evolution of DNA research. It does not include any guidance or discussion in theological and philosophical import. It could be humanity's warning shot across the bow for humanity, portending a diabolical or superior future for all of *Homo sapiens*.

Pharmaceuticals

Pharmaceuticals are a medical chemical, or drug, that enables a therapeutic change in the human body. They can be simple, like water, or complex molecules. They are rare. With the thousands of drugs in use, there are a very small number of targeted reactions within the human body. Of the millions of variant biological molecules in the human body, only about 250 or 0.025 percent are affected by our current pharmacological cornucopia of drugs.

A drug must only bind to biological molecules that are reactants to it and bind only to these molecules to work. Should they bind to other molecules, the drug would be a poison to the human body. Most molecules can barely achieve this level of discrimination. On the other hand, proteins have been designed specifically for this purpose. Proteins are the center of the biological world. They are the enablers, regulators, and gatekeepers of cellular reactions. Proteins are the switches that drugs seek to turn on and off, and thus are the most potent discriminating medicines in the pathological drug world.

To make a protein, you need its gene, and the recombinant DNA technology provides the missing link. The cloning of human genes allows scientists to manufacture proteins, which opens the possibility of targeting the millions of biochemical reactions in the human body. The protein and its manufacture mark the ability of scientists to introduce an amazing level of pharmaceutical drugs to benefit all of humanity.

Genentech started to produce a human growth hormone (HGH) in 1982, after its success with insulin. In 1986, they produced alfa interferon, a protein used to treat blood cancers. In 1987, they made recombinant TPA, a blood thinner to dissolve clots that occur during stroke or heart attack. In 1990, they started creating vaccines from recombinant genes. The first was against Hepatitis B. They sold a significant portion of their company in 1990 for $2.1 billion.

The gold rush has just begun.

Legal Matters

By the 1970s, it was possible to test babies for DNA, from sex to medical deficiencies. Suddenly the birth of babies was changing. Down syndrome dropped by almost 40 percent in the Western world. Other parts of the world chose to favor male babies. Aborting the fetus of female infants in those areas, prior to birth, became almost the norm in cultures where family size was regulated. Philosophical and theological discussion around this developed rapidly and exists to this day. Now in June 2019, the United States of America is debating abortion again in several states with the theological component being predominant. The example of China having a huge increase in males, and other cultures preferring male babies over female to enable inheritance of family resources, etc., has exacerbated this discussion.

A landmark judgment in the United States in 1977 stated, "Potential parents have the right to choose not to have a child when it can be reasonably established that the child would be deformed."[45] Commenting on this a reporter stated, "The court asserted that the right of a child to be born free of genetic anomalies is a fundamental right!" The double-edged sword of thought is that we also can use this to rectify birth defects in the womb or prepare ourselves for the birth of those with significant medical issues. This has not only allowed abortions but has also increased the ability of medicine to rectify problems within the womb or shortly after birth. I have a good friend who was known to have hydrocephalus before birth. His parents chose to have him regardless and what a fine example of a man he has become. The world is a much better place because Michael is in it.

---

[45] The New York Appellate Court ruled on December 11, 1977 in favor of Steven and Hetty Park and against Herbert Chessin for the wrongful life of the Parks' child. In a wrongful life case, a disabled or sometimes deceased child brings suit against a physician for failing to inform its parents of possible genetic defects, thereby causing harm to the child when born.

Eugenics was reborn, now popularly called "newgenics". Genes were now the units of selection, specifically the genes of the composition of the fetus. As late as 1996, the idea of abortions was raised in the court system, and just as recent as this year, we see protests both for and against abortions.

A sperm bank was created for the donation of sperm from men of the highest intellectual caliber, to be used only to inseminate healthy intelligent women. This bank was not successful, and the public spurned the idea. However, the concept of gene identity and gene health was accepted and entrenched in our everyday thinking: purification of the race by selective breeding.

Stop, Think, Discuss

In a culture where sexual activity has almost become a sporting event, with as little regard for morality as grass has for trees, and the often-resulting byproduct of pregnancy (if it is between opposite sexes), abortion has become a nonchalant byproduct. What is your opinion of this new morality? How would you enforce your morality (either side of the debate) on the general public, or should it be a concern for the public?

Question one: In these statements, does God have any opinion about abortion? Is God even involved? Should humans be allowed to abort a human fetus?

Question two: What is your opinion about developing a wide drug regimen based upon a protein? Do you see any negative consequences to having drugs that can affect every part of the human genome, or should we continue with only the 250 chemical elements that are currently in use in pharmaceuticals?

Question three: Was DNA alteration an idea created by God or by man?

Question four: Absolutely everything that happens has a precursor that makes it happen. Then the event happens. Could you have determined, before the discovery of the human genome, that science would unravel the DNA sequence of the human species?

Question five: Describe your understanding of the potentially good or bad in this science, in your own words.

Question six: What is it within this science that you don't understand? Do you think this science is too futuristic, like science fiction, to be true?

CHAPTER 22
The Human Genome Project

The enormous stretches of DNA between genes are called intergenic DNA, which has been compared to the long stretches of highway between cities. A gene is itself broken up into segments with long spacers called introns, interposed between protein coding segments. Intergenic DNA and introns do not encode any protein information. Some of the stretches encode the expression of the gene in time and space; they encode on and off switches appended to genes. Other stretches have no known function: "There were long stretches of DNA in between genes that didn't seem to be doing very much; some even referred to these as 'junk DNA', though a certain amount of hubris was required for anyone to call any part of the genome 'junk', given our level of ignorance"[46] (Francis S. Collins 2007). The structure of the human genome can be likened to speech that has pregnant pauses and interjections throughout the statement with each having a task in the sequences of words. An "on" switch codes a signal of when and where to activate a gene, and others code when to deactivate the gene. In 1992, J. Craig Venter, a Vietnam veteran who was considered tenacious but ordinary in the genetics world, started his own company, called TIGR, away from the world of university- and government-funded research in the search for the human genome. First, he decided to sequence the bacterium that caused lethal human pneumonia. Without going into the actual process, suffice it to be said that he was successful, and as a result was recognized for his magnificent work.

Successes in the bacteria field just hinted at what was to come. Meanwhile, the Human Genome Project continued its work in the dedicated fields of university laboratories and government facilities. Laboriously, the Human Genome Project

---

[46] Francis S. Collins, *The Language of God: A Scientist Presents Evidence for Belief*

slowly unwound DNA, particle by particle, never missing a step. The purpose was to create a pure science without skipping any steps of discovery. The scientists at TIGR went in another direction. They found comparable links and partial matches and jumped from gene to gene, only completing sequences when necessary. Theirs was a shotgun approach.

The conventional government and scientific approach required money, labor, and patience. MIT had a formidable team of scientists, mathematicians, chemists, engineers, and computer hackers. There were a dozen groups employed in other universities from around the world to assemble the data. Even so, with all these resources, the human genome was not predicted to be unwound until four years later.

Our Vietnam veteran found his association with TIGR to be disagreeable and went in another direction. He formed another company, called Celera. Celera announced a shotgun acceleration to solving the human genome in record time. The war veteran, Venter, agreed to make his findings a public resource except for his desire to seek patents on 300 of the most important genes that would act as targets for drugs for medicines such as schizophrenia, breast cancer, and diabetes. He said he would have the puzzle solved by 2001. This threw the public trust into frenzy, and the Government of the United States threw open the coffers to stay ahead of this upstart private company. Genome project scientists at the National Human Genome Research Institute in Maryland and the Sanger Centre in Cambridge, England were using a more laborious but more precise technique to map out the human genome. The decision had already been made that the data in GenBank from the Human Genome Project was going to be made available to the public, including Celera.

Celera combined the nine percent of the Human Genome Project's publicly available data with their own research. By continuing to validate their own research with the public project's data, Celera achieved a seven to eight times redundancy of the human genome sequence.

Celera used the largest civilian supercomputer and a shotgun technique to piece together the sequences that make up the double helix of DNA. Celera scientist Mark Adams developed the algorithm for the shotgun technique, and the company used it to map the genome of the fruit fly. The shotgun technique involves randomly chopping up DNA and then piecing it back together like a puzzle, a technique that is 99 percent accurate. Because Celera's shotgun approach was 99 percent accurate, they only mapped the genome a total of four times rather than the seven or eight recommended by GenBank.

Using this technique, human DNA was formally recognized as having been discovered and with a great fanfare, the Celera group and the Human Genome Project, along with the President of the United States and the Prime Minister of England announced the successful discovery of the human genome in February 2001.

> Earlier this week ... scientists announced the completion of a task that once seemed unimaginable. That is, the deciphering of the entire DNA sequence of the human genetic code. This amazing accomplishment is likely to affect the 21st century as profoundly as the invention of the computer or the splitting of the atom affected the 20th century. I believe that the 21st century will be the century of life sciences, and nothing makes that point more clearly than this momentous discovery. It will revolutionize medicine as we know it today.
>
> Ted Kennedy[47]

Understanding the human genome would be quite another matter.

Each human cell contains twenty-three pairs of chromosomes, for a total of forty-six. Twenty-two of these pairs, called autosomes, look the same in both males and females. The 23rd pair, the sex chromosomes, differ between males and female. All other upright-standing creatures—the apes, gorillas, chimpanzees, and orangutans—have 24 pairs. Several million years ago, the human genome departed from the ape genome, acquiring new mutations and variants over time. And gained a thumb. Life evolved, and it also answers the often-repeated statement: "If we evolved from apes, how come there are still apes?"

This is the story, the reality, the truth. It is the supplement to the ancient texts that try to explain our origins. It is not a coincidence, nor is it the final patented blueprint. This story is revealed in our development upon this Earth as we have need and ability to understand it. It is not only the effort of brilliant men, but the mystical, implanted desires of force within all of us to understand our meaning upon and within this universe.

---

[47] Ted Kennedy, senate session, Congressional record, June 29, 2000.

Stop, Think, Discuss

What is our purpose in this great cosmos, and why is this information revealed to us now?

If *Homo sapiens* did not exist, would the universe exist? if yes, what is our purpose and why are we a part of it?

Never stop questioning our knowledge or those who would interpret it to us. Every day, a new facet of our existence is revealed that significantly evolves the story as we knew it yesterday. Be on guard of those who tell us they have all the answers as they usually only have their own egos to reveal, and control is likely their only goal. Knowledge is really knowing that we must continue to explore and question everything. Ultimately, everything will be revealed.

By the same token, be aware when you are being told the truth, for the truth will give you peace of mind and comfort for your soul.

What is your opinion of the science of evolving in nature? Can this agree with theological statements?

Question one: Did you know that the ape kingdom has twenty-four pairs of chromosomes, while *Homo sapiens* have 23?

Question two: When you read that a business manager is creating the human genome for business profits, how did you react? Will this endeavor be concerned about the safety and wellbeing of man or concerned about profit?

Question three: What is it that drives us to understand the mystical forces that drive civilization and our roles within it?

Question four: Absolutely everything that happens has a precursor that makes it happen. Ancient texts tried to give explanations about our species, and this science is an update, with complete revocation of ancient understanding. What is your opinion about abandoning these ancient understandings, usually from a theological perspective?

Question five: Give your understanding of the dialog within this chapter. How does it affect you?

Question six: Is this discussion correct in its entirety or does some nonsense exist that you would like to explain?

CHAPTER 23

DNA Tests the Atheist Mind

Vipers and biters, stingers and pests,
or philosophers, scientists, theologians which test,
with hooligans, robbers, and liars and cheats,
Which is the one that that triumphantly beats,
us into submission and rules the day.
The atheist way?

Author unknown

DNA investigations solving the human genome led one atheist philosopher to assert that there is a creator. He is living proof that investigations with an open mind allow one to change direction in thinking upon demonstration of facts enough to present the need to change direction. This book is about reasoning with an open mind, hopefully with others, and reaching a conclusion that satisfies your personal ideology with clear thinking and facts.

Antony Garrard Newton Flew (1923–2010) trained as an English philosopher. He was analytical and Darwinian in his thinking. His notoriety was for his work in public discussion of the philosophy of religion, much like Richard Dawkins of the present. Anthony Flew taught at the universities of Oxford, Aberdeen, Keele, and Reading, and at York University in Toronto. It was after retirement and during his analytical (mathematical and statistical) thinking of the relatively new science of DNA that he became aware that there had to be a creator in the evolution of life. He never adopted a specific religion, just a deist point of view. There is a creator

God, all-powerful and perfect, overseeing everything out there and watching, but not participating in everyday living.

Flew was a strong advocate of atheism, stating atheism was the foundational premise we should follow until there was proof for God. In basic terms, only when you touch, taste, see, or feel that hand of God should you believe in a God. In 2004, Flew stated that he had come to understand that his lifelong commitment to follow evidence, showed the existence of a God. Additionally, he was most impressed by the evidence for Christianity.

So, what changed? The main reason, Flew stated, was recent scientific work on DNA, which points to the origin of life through involvement of a "creative intelligence". His change of thinking, and in keeping with his philosophical and analytical thinking, was nearly entirely due to the complex DNA discoveries that overshadow the statistical impossibility of chance. As he states:

> What I think the DNA material has done is that it has shown, by the almost unbelievable complexity of the arrangements which are needed to produce (life), that intelligence must have been involved in getting these extraordinarily diverse elements to work together. It's the enormous complexity of the number of elements and the enormous subtlety of the ways they work together. The meeting of these two parts at the right time by chance is simply minute. It is all a matter of the enormous complexity by which the results were achieved, which looked to me like the work of intelligence.

> Why do I believe this, given that I expounded and defended atheism for more than a half century? The short answer is this: This is the world picture, as I see it, that has emerged from modern science. Science spotlights three dimensions of nature that point to God. The first is the fact that nature obeys laws. The second is the dimension of life, of intelligently organized and purpose-driven beings, which arose from matter. The third is the very existence of nature.

Anthony Flew[48]

---

[48] Anthony Flew, from his book *There is a God,* published 2007

Anthony Flew was in the camp of "show me" if you want me to believe what you are saying. He was the predecessor of Richard Dawkins and just as controversial as Dawkins is today. Using his own skills and reasoning, he determined that the possibility of life creating itself through coincidence was not even remotely possible.

The current candidate perplexing theologians is Richard Dawkins. Born in March 1941, he was the University of Oxford Professor for Public Understanding of Science from 1995 until 2008. His field was evolutionary biology, which included the study of DNA, and he was highly regarded and prolific in that field. He loves the attention of radio and television, especially the attention given when he speaks on atheism, and he does everything possible to provoke those who have opinions other than his. This has become such a part of his life that one almost must question what his motives are. There is a lot of recognition, and there is a lot of literature sold under his moniker. He also has an adoring audience who disdains anything religious or anything suggesting he is not perfect. He has an almost God-like persona. But then, in killing God, is it possible that he is likening himself to a god or a replacement for God?

I was recently able to attend a conference on theology and science at the University of Minnesota. I watched as several hundred ordained theologians aggressively perked up, with blood pressure rising, when Dawkins' name was introduced in discussions. It gave new meaning to the beginning of an old joke I know, which originally was about tight shirt collars: "Their heads pounded, their eyes bulged, and their breath came in short gasps …" So, I decided to listen to some of the Dawkins discussions that have been recorded.

My attention in these recordings is that Dawkins has credibility in his field of science and has a peer group proving his science, just like any other recognized scientific investigation. When he diverted to theology, he left reason and proofs behind. I heard "Trumpian" explanations of his opinion of religion. He has created, in his own mind, a religious fervor of a non-existing religious Christian cult. His description of Christianity is so off the wall as to be less than simplistic, even to an untrained theologian such as myself. Baiting, aggressive opinion, unrealistic history, and lack of basic knowledge of Christianity define his arguments for atheism. His description of his thinking on this subject is far less than ontology or free from personal bias. As a result, he is popular, sells a lot of literature, and is recognized by a huge number of people who are looking for a leader or modern-day hero. Perhaps a champion of mysticism, as advanced by modern literature and cinema, extoling superior human or human-like abilities?

Stop, Think, Discuss

I have deliberately mentioned Christianity here in this atheist discussion because I do not find any negative discussion from atheists about Islam, Sikhism, or Hinduism, etc., in the verbal grenades tossed into the public arena of opinion. Perhaps the thought in the back of their minds is of Salman Rushdie and the fatwa against him, calling for his assassination, which was issued by Ayatollah Ruhollah Khomeini, the Supreme Leader of the Islamic Republic of Iran? Christianity is the soft target, with its foundational tenants of love, forgiveness, and grace. There are no death threats, fatwa, or jihad in Christianity.

Can you give any scientific reasoning that gives clear direction and understanding for the ideals of morality, love, ethics, or answering the question "why"?

Question one: Anthony Flew was the bane of the study of theology at one time. Philosophical reasoning occupied the time of most theologians, certainly those in studies toward a degree in that field. What caused Flew to change his mind, and was this, perhaps, a revelation from God? Is God even involved? Is this God's idea or one of Flews' own, made without outside influence?

Question two: Flew started from a premise that we should not consider the work of any god until it is proven that a god was involved. What is the *ontology* of his thinking? That is, what is the actual information, free from our personal bias, of his premise? And what is the *epistemology*? That is, what is the knowledge, as learned by him, including his preconceived notions, of his statements?

Question three: Apparently, Flew worked to his resolution backward from the norm in his way of deciding about God. Was this a way for God to help him understand, in a procedural way, that gave him public resolution to the actual truth?

Question four: Absolutely everything that happens has a precursor that makes it happen. Then the event happens. What do you think caused Flew to go forward in a manner that was contrary to the public way of accepting God, before proving it publicly? Also, could Flew have taken a more normal way to his truth if he had prior faith in the existence of God before he started?

Question five: How would you compare Flew to Richard Dawkins, in your own words?

Question six: Are there issues in these statements with which you disagree? Bring these issues into a discussion with others.

PART FOUR

The Sun, Moon, Stars, and Everything That Shines in the Sky

What your children know about the stars and you don't.
Chris Pedersen

To learn, one must not only read but must also think.
Thinking is best done through discussion.
Chris Pedersen

Stop anywhere it would be prudent to ask any question.

CHAPTER 24

Feel the Science as You Examine It

I start this volume with my own experience, and I ask your indulgence. It is what I have experienced with my own being. My purpose of this is to relate what others have discovered and are telling us in their own way. I should start with the big bang, but I feel I need to give my own experience as prior thought to bring relevance to all the thoughts of others.

This is what I have seen with my own eyes. I am an aviator. I fly. I have climbed to great heights and seen things you groundlings have never imagined. I have been in rarefied air where a lack of oxygen for fifteen seconds means certain death, where a loss of pressure would cause your blood to boil. The sun, moon, and stars have risen at my feet, and I have felt dwarfed by it all. At the same time, I have felt magnificent, with the opportunity to see all the wonder of the creation of the universe right there in front of my eyes. The pure sky, without pollution. Stars that just beg me to reach out and grasp their being. A Milky Way that is as bright as a map.

I have risen on still air, in a craft without motor, to the very edges of space, witnessed giant mountain ranges small and quaint, as if painted on a canvas, and observed great oceans of green, verdant and pure. I have chased the curve of the Earth as we rolled around the sun and looked for hundreds of miles across a curved horizon like the edge of a ball; the air so cold you would freeze solidly in an instant if you were outside your sturdy ark.

I have watched the moon as the Earth's shadow quieted it into a full eclipse and wondered at the morning sun rising at my feet and larger than anything I have ever seen while glued to the Earth. I have seen the aurora dancing all around like a ballet choreographed in color and then quickly rushing with the speed of light to hide from the rising sun, but undulating in ever rapidly growing movement as it went—crystals of water trapped in outer space, sent there from furious storms that spent their energy in tantrums of fright.

The tops of great storms have reached out for me, to grasp and shake me, like a dog shaking a slipper, with lightning and thunder all around as I contemplated the smallness of my being, and great storms and wind so strong that nothing could stand in its way. Through it all I flew, undeterred, confident, and with eyes of wonder. Through misty cloud that captured the dying rays of latent sunlight in the spectrum of magenta—the whole sky turned pink, up, down, forward, and back. I saw a parallel universe through the eye of a hurricane as it passed over the airport where I had just been, and the amazing sight of it from above as I reached out for safety in the height of space.

If I composed a dissertation from my pinnacle above the sky and described what I saw with my own eyes, you would not believe it. If I told you it was a dialog from outer space, on a journey to another place, you would hang your jaw in awe. Your dreams would turn to a desire to go, see, wonder, and contemplate the vastness of it all and our place within it.

I have journeyed back and kissed the Earth, like the wings of a butterfly caressing our skin, renewed with understanding and the confirmation of knowing my place, my joy, and my purpose upon it.

But it is yours. Look up. All you need to do is open your eyes to see. Listen to your heart to appreciate, and let your mind absorb what you are looking at. The sun, the moon, the stars, and the Earth below. All there. All yours. All of it, beyond your imagination and yet real to the smallest detail. Believe.

Before we go back to the beginning, the big bang, let's look at what historical *Homo sapiens* had to say about our universe. In terms of timeline, we are infants to this planet. We can safely date our existence in present form back two hundred thousand years. Our cognitive ability can be safely dated back to eighty thousand years. With cognition, our skills developed into clear thinking and language, then agriculture about ten thousand years ago. Written language demonstrates a level of superiority over other species on this planet, and Sapiens are the only species to cook their food, eliminating bacteria and other harmful disease, thus increasing our life span.

Thoughtful dissertation and clever insight are recorded with the Greek philosophers and scholars about three thousand years ago. Other cultures in the East were also pondering scientific expression at the same time, but that discussion of compatibility and time is for another day. Those early Greeks were the first to suggest that the Earth revolves around the sun. Democritus was the proponent of atomization. He suggested that everything is made of minute particles of matter, which he called atoms. This was in 460 BCE. How is this possible? Other cultures were rubbing two sticks together to make fire while the Greek philosophers were dissecting the anatomy of the stars.

Religion also has a major role in our understanding of the universe. Popularized metaphors about the creation story have permeated our religious cultures through religious text. In addition, the ego of man as centrist, as the potentate of all being, has influenced our historical review.

When religion and narcissism conflict with science, philosophy enters the discussion. This makes a stew that exists to this day. Since religious belief is waning, and spirituality is growing, there is a populist belief that science has all the correct answers. Sadly, this also removes the original mysticism of basic religious expression and places it outside rational thought for many. Today there is a three-ring circus: atheism, focused scientific reasoning, and people who just follow popular issues with vacuous abandon, believing that only science has the answers necessary to understand the existence of our being. Atheism in scientific expression has become quite a normal state of thought, and those who have never allowed their minds to have original expression jump on the scientific bandwagon with glee.

Philosophical thought could suggest that we *Homo sapiens* rethink some of the origins of the universe. Science is correct as far as science understands in the moment, but science is always changing with further investigation. Every day, new discovery is collated and disseminated to our species. Historically, science can easily take us mathematically to 380 million years after the big bang. Mathematics doesn't lie. It is the universal language. But mathematics does not tell the whole story. What happened one millisecond prior to the big bang? What was the cause? We know the effect. In a global sense, it is us. But why, and what caused all this extravaganza? An atheist point of view is that it is coincidence, pure and simple. A theistic point of view is that it is mystical. The mysticism of religion has been lost in all the scientific reasoning, and perhaps this is an error of civilization in present times. The failure of theistic thinking is the inability to realign with the known facts as they exist in the present and include metaphor to account for the ancient teachings that instructed morality and ethics to the uneducated at the time.

There is a place for theistic thinking, remembering that a belief in God is the basic building block of *Homo sapiens* civilization from the beginning. A leader in the scientific community, John Polkinghorne KBE FRS,[49] is a renowned English theoretical physicist, theologian, writer, and Anglican priest. He does not stand alone with argument for a God in science, and publicly debates those within science who would say otherwise.

---

[49] John Charlton Polkinghorne KBE FRS is an English theoretical physicist, theologian, writer, and Anglican priest. A prominent and leading voice explaining the relationship between science and religion, he was professor of mathematical physics at the University of Cambridge from 1968 to 1979, when he resigned his chair to study for the priesthood, becoming an ordained Anglican priest in 1982.

Stop, Think, Discuss

The Babylonians created astronomy in the belief that God was in the stars and was speaking to us through the stars. Although your daily horoscope has science-like qualities in reasoning, to accept this prescription of communication, you need faith. The Hindu faith is completely engulfed in Astronomy on a day-to-day basis, for instance.

If you need faith to accept the value judgment provided, is it science or religion?

When you discuss this objectively, are you in philosophical discussion, or trying to persuade others to your point of view?

Question one: Have you seen the moon rise from below you, under your feet? Have you seen it twice as large as when it is high in the sky? When you see something like this with your own eyes, can you wonder what created all this without God's influence even being involved in this creation? Is this God's creation or man's?

Question two: What is the *ontology* of these statements? That is, what is the actual information, free from our personal bias, of these statements? With Flat Earth societies springing up all over the Internet, can you understand what is going on in their minds? Are these Flat Earth folks just fooling with the Internet, or do they not believe what their eyes are providing as evidence? For some, is their religious dogma so powerful that they do not believe what their eyes are telling them?

Question three: Can you understand why ancient civilizations believed in a flat Earth? Was it an understanding created by God or by man? Explain.

Question four: Fundamentalist Christians believe in a literalist interpretation of the Bible. Therefore, the big bang is anathema to them. Other theologies have much different interpretations, but much apart from the scientific reality. How can we reconcile these points of view with proven and visible realities of evolution and the big bang?

Question five: How would you describe your view of the sky, in your own words?

Question six: Are there issues in this chapter with which you disagree? Give your explanation.

CHAPTER 25
The Big Bang

What your children have been taught and you should know, depending upon where they went to school, is that: The universe is more than 13.8 billion years old. It started with an event called the big bang, which was not an explosion at all. Rather, 13.8 billion years and one second ago, the universe was plasma, liquid, molten matter. It was in that state, at around 100 million, trillion, trillion degrees Celsius, like a sea of lava, but a sea that was only the size of a pinhead in comparison to the whole universe. In the one second prior to 13.8 billion years ago, something cataclysmic happened to cause all that molten matter to coalesce.

To imagine the process, think of a megaphone. You speak into the mouthpiece, which is then amplified and expelled into the atmosphere with a much larger volume and which spreads as far as there is an atmosphere and waves that do not deteriorate in the atmosphere. This is the big bang. The glob of plasma was cataclysmically engaged, like a bat hitting a ball, and the matter immediately moved in a method like the megaphone described above, spreading out and moving forward into this vast empty universe.

The plasma immediately started to cool. This cataclysmic event is measured in Planck time. Max Planck,[50] the scientist, thought in units of time as one 10 million, trillion, trillion trillionths of a second. Consider time this way: a nanosecond, 10^9, which is much longer than a Planck 10^{44}, indicates that there are as many nanoseconds in one second as there are seconds in thirty years. WOW! Using Planck

---

[50] https://en.wikipedia.org, Max Karl Ernst Ludwig Planck (1858–1947) was a German theoretical physicist whose discovery of energy quanta won him the Nobel Prize in Physics in 1918. Planck made many contributions to theoretical physics, but his fame as a physicist rests primarily on his role as the originator of quantum theory, which revolutionized human understanding of atomic and subatomic processes. Planck came from a traditional, intellectual family. His paternal great-grandfather and grandfather were both theology professors in Göttingen.

time, scientists mathematically measure the temperature, molecules, and rays of the cooling plasma as it coalesced into neutrons, protons, and then atoms, and continued into balls due to gravity. Contemplate this quotation before returning to the big bang:

Science cannot solve the ultimate mystery of nature. And that is because, in the last analysis, we ourselves are a part of the mystery that we are trying to solve.

Max Planck

Large globs of matter formed, creating planets, while yet other globs, with nuclear fusion involved, became stars. Each of the bodies gathered, one to the other, and became constellations. Their spinning action made them round, ball-like. Gravity held different lumps together, and the constellations had individual stars within them. The stars gathered planets and moons around them, orbiting them as we see through the telescopes today. We only see about 2.5 percent of the stars through the telescopes, even using the Hubble telescope orbiting Earth in space. The rest is out there, mathematically significant and guaranteed, but unseen.

The measurement of gravity in mathematics indicates that in space all the organized structures we see, and calculate, cannot happen unless there is significantly more material out there than we can currently see. Something very large, in terms of gravity, is in space, but can only be identified as "dark matter", known but unseen. Even more important is a force that is massive, controlling the movement of the constellations away from each other into deeper space, but which can only be identified as dark energy. More on that later.

The constellations move randomly to other constellations, each as a unit within itself, defining that each constellation is independent from the others. All the stars and planets move in relation to each other within all the constellations.

Pretty straightforward stuff today, but in history, not so much. Today, in most developed areas where we have streetlights, most people hardly even look at the sky, and if they do, it is obscured by the ambient light of the streetlights and so on. In areas and time before electricity, everyone knew the sky. It was a central part of their being.

No man's knowledge here can go beyond his experience.

John Locke

Stop, Think, Discuss

If we can potentially only see 5% of the whole universe, but we currently see only 2.5% of that 5%, what visible and/or significant role does *Homo sapiens* play in the drama of the whole universe?

If we currently understand that 95% of the universe is there, somehow, but we have absolutely no intimate knowledge of it otherwise, could we possibly have enough knowledge to know one way or another about a divine hand in creation? If yes, what process is it that makes us superior or lords of the universe?

Question one: In these statements, what is the divine creation of God and what is the descriptive hand of man? Is God even involved? Does this description of the big bang reflect God's creation, or is it an interpretive and imaginative description created by man?

Question two: The universe is 13.8 billion years old. The planet Earth is an insignificant planet with one moon within the constellation Milky Way, which is in turn insignificant in the universe. *Homo sapiens* became a life form approximately two hundred thousand years ago in this sea of insignificance. What trick of mind makes *Homo sapiens* think they are the center of all understanding of the whole universe?

Question three: What came before the big bang? Was the big bang created by God or by something else? Or was it a coincidence?

Question four: Absolutely everything that happens has a precursor that makes it happen. Then the event happens. Regarding the big bang, can you determine, before the event, the cause of the event? Also, could the event have been altered if the cause had been known beforehand?

Question five: How would you write this scientific theory, in any Bible, in your own words, if you were asked to write it for posterity?

Question six: Are there issues in these statements with which you disagree? Has Harry Potter— creative thinking—taken over the dialog on this issue?

CHAPTER 26

The Stars Give Direction

Astrology

The Babylonians, like Darwin with species, were cataloging the stars. They developed astrology. To them, the planets were created by the gods. So, they developed detailed records over history to interpret communications between the celestial (the gods) and terrestrial (us). It wasn't until the mid-seventeenth century that pure astronomy diverged from astrology and the perceived communication between God and man.

Over time, astrology developed into a much stronger belief, one that carefully predicted the movements of celestial bodies and studied their relative positions in the sky. Based on that, practitioners of astrology would be able to divine events that would occur on Earth, both to themselves and to others. Astrology is most closely linked to horoscopes, or reading about one's daily/weekly destiny, as well as one's fundamental personality, based on the star sign under which one was born.

For thousands of years, astrology was considered a legitimate scientific field and was spoken of with similar respect as astronomy, geology, and the other natural sciences. Despite the lack of experimental proof, and requiring only faith to accept, this powerful tradition existed all over the world and was simply accepted as a part of the human experience. It is still highly acceptable by some and used daily in the practice of the Hindu faith for instance.

While millions of people around the world still read their horoscopes every day, and thousands of career practitioners earn their living reading star charts and tarot cards, there is no scientific basis for astrology. The only true power of this area of study is the power given to it by people who are eager to make some order out of life's occasional chaos. It is a faith-based system of belief, using observed criteria to create a sense of reality.

Astronomy

Like astrology, astronomy has a beginning that dates back thousands of years, which is often why the two areas of study are mixed up or assumed to be the same. Astronomy is the study of celestial objects and phenomena through the application of physics, chemistry, and mathematics. These disciplines help to shed light on the origins of these objects, their composition, and their manner of interaction.

On a basic level, the term astronomy is a catchall for the study of anything that lies outside the atmosphere of Earth, ranging from the moons of Jupiter and the comets of the Kuiper Belt to the farthest-flung galaxies, invisible black holes, the big bang theory, and the cosmic microwave background. Most modern studies of astronomy fall within the purview of astrophysics, but within this popular field, there are also theoretical and observational branches. These two areas work in conjunction, as the creation of computer models and intensive analysis of data is critical to supporting observations of distant galaxies, stars, comets, moons, supernovae, and other phenomena.

For a vast number of people, Astronomy is one of the most widely respected and exciting scientific fields of human perspective, as it has not only taught *Homo sapiens* about the fascinating contents of our solar system, galaxy, and universe, but also more about life on our planet, as well as its origin. So, the more we know about what happens out there, the more we understand what's going on inside us, even at the microscopic quantum level. Prior to the 1940s, popular science was metallurgical, chemical, or medicine in history—turning lead into gold, for instance, the Midas touch; Leonardo da Vinci studying anatomy; or using leeches to cure ailments.

Unlike astrological gurus and "experts", astronomers rely on the strict and proven scientific method to develop their theories and test their predictions, making astronomy a legitimate and invaluable field.

Astronomy and Navigation

Society regulates itself around three natural astronomical celestial clocks. The daily rotation of the Earth, with the apparent rising and setting of the sun, gives us the day. The rotation of the Earth around the sun gives the year, and the moon's circuit around the Earth for moon phases. The unit of the hour is exactly the movement of the sun in the sky by fifteen degrees per hour. If you have asked why the clock is based upon the 60-base rule, it is all because of the mathematical composition of the vectors of the position of the sun. Our planet rotates fifteen degrees per hour, three hundred and sixty degrees make twenty-four hours. Look at the position of the sun right now and determine how many degrees it must go to sunset. Every fifteen degrees you observe is the number of hours to darkness. Break it all down, and you get a lot of 60-base components, just like dividing a circle.

This, in turn, is also a navigation device, as well as the position of the stars at night. If you want to determine the direction of south in the Northern Hemisphere during the day, take your wristwatch and point the hour hand directly at the sun. Sorry, you are going to have to use your imagination if you are wearing a digital watch. The halfway position to the twelve o'clock position on the face is south. Adjust this for daylight-savings time if you are doing it for a lifesaving exercise in the wilderness. In the Southern Hemisphere, the resultant is north.

This is all history. In 1675, King Charles II commissioned the Royal Observatory in Greenwich, England for the purpose of perfecting the art of navigation. Using this information and a good timepiece, a sailor could determine his position on the surface of the Earth. By approximately 1975, this art of measurement of the stars was usurped by inertial navigation in modern-day airliners. This was developed by several manufacturers and solely owned by them. A leased mechanical device is placed upon an aircraft that within its small structure emits a light through a gyroscopic prism that breaks the light into two beams. This spectrum is cast upon a medium that can measure infinitesimal movement of the light. With accelerometers and calculations all contained within the unit, it can tell you the distance from its starting datum, which is set before departure. This unit is accurate to within feet of the true desired track on its intended path or altitude.

Then, satellites were launched by the United States government, creating a network of global positioning satellites around the world. By use of a ground-based mechanical device, the electronic waves produced by these satellites can be measured and position can be calculated without reference to the celestial bodies above, and in a much more economical way than inertial navigation. Now, most advanced countries have developed their own GPS satellite system, so they are not dependent on the United States, who have built the ability within the system to fault the system in case of attack upon their country. This would cause missiles to target something other than the accurate coordinates so conveniently provided to them by the USA.

But the stars are used for much more than navigation. In addition to navigation, the use of stars as a coordinate for building structures is also a field of astronomy.

For at least ten thousand years, *Homo sapiens* plotted the movement of the stars, constructing calendars, creating astrology, and erecting monuments to track the parade of constellations, comets, and planets across the night sky.

Like the Babylonians, many ancient cultures shared a belief that we came from the stars. Perhaps this explains all the time, effort, and energy that was put into the study of the heavens, from then to today.

Several constellations and stars played a significant role in mythology about the origin of our species. In places where these legends emerge, structures exist that precisely correlate their position with these stars and constellations in the sky. Orion is the most prominent of these.

The Importance of Orion

The constellation Orion is the most recognizable constellation in the night sky, and its location on the equator allows it to be seen everywhere on the planet. Orion was central to many ancient cultures.

We know from the Pyramid Texts, which are among the oldest religious writings in the world, that the ancient Egyptians believed that the gods descended in the form of human beings from the belt of Orion and from Sirius, the brightest star in the sky, following Orion over the horizon. This is critically important in Egyptian cosmology. Orion was associated with the god Osiris, and Sirius was associated with the goddess Isis, who together are said to have created the whole of human civilization.

To locate Sirius in the sky, follow the three stars in the belt of Orion down to the horizon. The brightest star next to come up on the horizon is Sirius. Not only is it bright, but it also blinks white, green, and red as it rises in the night sky. During the Second World War, young inexperienced allied fighter pilots flying night patrol over the desert in the Middle East or Africa would see this and think of it as the navigation lights of the enemy and often fly evasive patterns to shake off the "fighter pilot" on their tails.

Alignment with the position of the stars in Orion, especially its belt, matches precisely to various ancient structures, from the pyramids at Giza to those found at Teotihuacán in Central Mexico.

The magnificent pyramids at Giza are one of the ancient wonders of the world. Together they enunciate the concept of sacred alignment, as they are a precise three-dimensional map of the stars in the belt of Orion. Additionally, their size and placement also reflect the brightness of the stars in Orion's belt. There are also those who believe that the Giza plateau is the geographical center of Earth, and the central pyramid, Khufu, is more aligned to true north than is the Greenwich observatory in London.

The ruins of the ancient city of Teotihuacán lie in the highlands of Central Mexico. This is another marvel of the ancient world that is connected with constellations. Teotihuacán was one of the largest cities in the world during its lifetime, having a population of approximately 150,000 to 200,000 people. They called it Teotihuacán, meaning the "Place of the God", because they believed that it was the place where the current world was created.

Like other sacred sites, the observatories, pyramids, and structures at Teotihuacán mirror star alignments. The complex at Teotihuacán contains three pyramids, two larger and one smaller, boasting a similarity to the layout of the pyramids at Giza and duplicating the belt of Orion. These pyramids are known as the Pyramid of the Moon (furthest north), the Pyramid of the Sun (central), and the Pyramid of Quetzalcoatl. The Pyramid of the Sun is said to be aligned with the Pleiades, another constellation of great importance in myth and lore that is often connected to the constellation of Orion. The alignment clearly reflects the alignment of the three stars of Orion's belt.

The Hopi, a Native tribe in North American, also have monuments modeling celestial bodies, and the landscape is purported to have connection with the constellation of Orion. They live in an area consisting of three mesas in northeastern Arizona and have been there for more than a thousand years.

They say the natural structure of the three mesas mirrors the three stars in the belt of Orion, and their history states that this is why the Hopi settled in this location. They believe this place to be the center of their universe, where contact can be made with the gods. Some say that when connected to other Hopi monuments and landmarks around the southwest, the collective sites are supposed to map the entire constellation of Orion. I will leave this up to your imagination.

Orion is obvious, and it became important. It is possible to mathematically compute that the positions of the stars of Orion have moved from those used by the ancients.

> *If you would be a real seeker after truth, it is necessary that at least once in your life you doubt, as far as possible, all things.*

René Descartes[51]

---

[51] René Descartes was a French philosopher, mathematician, and scientist. A native of the Kingdom of France, Descartes is widely regarded as one of the founders of modern philosophy. https://en.wikipedia.org

Stop, Think, Discuss

In Christian writings, there is the star of Bethlehem, which three wise men followed to find the baby Jesus in a manger. What do you think is the background to this story? Is it made up or mythical? Is it a true course of events, as described by the Bible? Could it have been a comet, the passing of which happened at the same time as the birth of Jesus, which resulted in a metaphor? How could three wise men accurately plot the exact alignment with the stable in Bethlehem?

Question one: In celestial prognostication all through the ages, was there a divine intention of God without the input of man? Is God even involved? Is the process of location of structures influenced by God's or man's creation?

Question two: What is the basic intention of man through the creation of structure by God's design, reflecting the stars?

Question three: How do you understand astrology in today's world? Is it science or theology or mindless fortune telling?

Question four: Absolutely everything that happens has a precursor that makes it happen. Then the event happens. What is the precursor to the ancient empires deciding that their cities and location of structures were God-inspired? What was their idea of God?

Question five: How would you describe astrology, in your own words?

Question six: Are there issues in these statements with which you disagree? Is some of this chapter creative thinking? Explain.

CHAPTER 27

When Religion Thinks it is Science

The most powerful religion of all time was given great power when Constantine saw opportunity to absorb the Jewish Jesus movement by melding it into a state power of centrist thinking in 315 AD. It was evangelistic—the only religion to endorse others from other cultures, and even women, into its core. With the power of state, it was given structure and the force of law. Armies were the enforcers of this new movement. It withstood the change of time for two thousand years with this man-made organization and corporate narcissistic personality. It was given the power of life and death. Those who stood in the way found, to their sorrow, what the power of the leaders of this enormous cult could do.

By the authority of God, who somehow was converted into emperor or pope, it was decreed that whatever or whomever disagreed with, or contrived against, Christians or even their thoughts, would perish. Thus, some of the great schools of Greek scholars were destroyed, if they even had the appearance of disagreeing with Christian authority. Books were burned, buildings torn down, and scholars spurned or put to death. The authority at that time believed in a flat Earth, and that the sun revolved around the Earth. That is what they saw with their eyes and that is what the ancient texts of the Jewish and now Christian Bible said. Power and might and glory to the potentates, and down with the contrarian. Believe or die!

The second most powerful religion, Islam, through Muhammad evolved during his lifespan around 600 AD. He—having been raised around Christian principles, yet a Muslim in faith—learned all too well the power and structure of state faith. Islam took control of the Middle East and North Africa and chased the Christians out, but with the same self-centered need for power, glory, and enforcement of personal grandiose self-certification. In addition to being political, the Islamic movement pro-duced the leaders of scientific thought until nearly the fourteenth century, whereupon the Christian movement, headed by Pope Innocent IV, granted their theologians the ability to think outside the confines of faith. Islam reverted to considering their faith and prophet more than science, and Christians started to allow scientific thinking, through Albertus Magnus[52] and Thomas Aquinas.[53]This all makes present-day scientific and atheistic thinking a very likely alternative to some of those who have focused on the academic process. Historically, pure science cares not about power, politics, or glory. Furthermore, science insists that everyday scientific changes and new revelations create a new and exciting future thought. Theological thinkers gener-ally do not have an open mind as to future thought, as this would modify their his-torical theological bias. But there are notable exceptions, like Sir John Polkinghorne KBE FRS, known for his ground-breaking work in bringing science and theology together. He is a British mathematical physicist and Anglican priest. KBE is Knight Commander of the British Empire and FRS is Fellow of the Royal Society—both extremely prestigious awards of recognition of service and accomplishment.

While religion would like to maintain the status quo, with exceptions, and rep-resent a belief system established when religion used its power to present itself as science, scientists, with exceptions, would admonish everyone:"Never trust the white coats (religious or scientific) who say they have all the answers to all the questions," because not even all the questions are known yet. An atheist would tell you they only care for themselves and not for political structure or the power of potentates. Religious thinkers need to dissect their perceived history of their belief system and rebuild it into something they can understand and accept without the baggage of historical and man-made hierarchy. The mystical approach of religion has a place in our thinking and has had since the creation of *Homo sapiens* more than two hundred thousand years ago. We just need to eliminate the self-serving errors of process. Sadly,

---

[52] Albertus Magnus, OP, also known as Saint Albert the Great and Albert of Cologne, was a German Catholic Dominican friar and bishop. Later canonized as a Catholic saint, he was known during his lifetime as Doctor universalis and Doctor expertus and, late in his life, the sobriquet Magnus was appended to his name. Scholars have referred to him as the greatest German philosopher and theologian of the Middle Ages. The Catholic Church distinguishes him as one of the 36 Doctors of the Church. https://en.wikipedia.org

[53] Saint Thomas Aquinas OP was an Italian Dominican friar, philosopher, Catholic priest, and Doctor of the Church. He is an immensely influential philosopher, theologian, and jurist in the tradition of scholasticism, within which he is also known as the Doctor Angelicus and the Doctor Communis. https://en.wikipedia.org

many students of theology can take apart the structure of religion, but sometimes get lost building it back together in a meaningful way with today's knowledge of science as it relates to our theological past. That leads to the condition I titled this chapter with: "when religion thinks it is science".

> *A little philosophy inclineth man's mind to atheism; but depth in philosophy bringeth men's minds about to religion.*

Sir Francis Bacon[54]

Then we encounter the modern-day flat Earth believers, who are hard to understand as they deny what most of the rest of us can see with our own eyes. If you Google this topic, you find thousands of educated people who currently support this and who come to this idea without provocation of outside influences. Then there are those who insist that their perceived fundamentalist or extremist religion agrees with science as it was perceived 1,500 years ago even prior to the dark ages. Their dissertations are that religions agree with their ancient point of view of religious interpretation of ancient documents as living works, not today's metaphors for the education of the uneducated at the time. This example given is Islamist, but it is only reflective of other religious beliefs of other religions that harbour similar concepts of the universe. For instance, Hindu ancient belief would say that the Earth started as an egg. Early Christian belief is documented in the Jewish and Christian literature as a "God-created the earth story over seven days" as described in the first and second chapters of Genesis, the first book of the Christian Bible and Torah.

Similar thinking can be found within Islam. I recently read a news article originally published in the *Gulf News*, April 10, 2017. This article pointed to the fact that an Islamic PhD student in Egypt had submitted her thesis for review on the hypothesis that the Earth was flat, unmoving, only 13,500 years old, and the centre of the universe. Further, she rejected the physics of Einstein and Newton, the astronomy of Copernicus and Kepler (which by inference would include all modern astrophysicists), the big bang theory, models of geological and atmospheric functioning, and almost all modern meteorology. The intent of this five-year study was to indicate that the books of Islam were correct in their interpretation of science when written

---

[54] Francis Bacon, 1st Viscount St Alban, PC QC was an English philosopher and statesman who served as Attorney General and as Lord Chancellor of England. His works are credited with developing the scientific method and remained influential through the scientific revolution. https://en.wikipedia.org

around 650 AD. Remember most branches within the Islamic tradition are constant in interpretation of their literature, always referring to the original documents to resolve conflicts within their theology.

This thesis was accepted by the university, represented by two assessors holding a professor rank, which represents the first stage of acceptance for granting a PhD in the chosen field. The president of the Tunisian Astronomical Association heard of this and immediately checked that this was not fake news or a hoax and set forth to do what he could to rectify this situation. He read the closing of this thesis and found the following conclusions therein. The results include: the Earth is flat and young, and it stands immobile at the center of the universe, which is made of only one galaxy; the sun's diameter is 1,135 km (not 1.4 million km), the moon is 908 km wide, and they lie 687 and 23 times closer to Earth, respectively; there are eleven planets; stars are "limited" in number and have a diameter of 292 km (not millions of km).

Clearly this PhD presentation was an attempt to create documentation that adhered to the historical documented discourse of Islam in or about 650 AD. Sadly, this is the best modern example of how theology has considered itself a science based upon ancient theological texts. Historically, there is also Christian-based theology that misled the world for centuries in many science discoveries. In approximately 1100, the Arabic leaders of the scientific world in Bagdad, the leading protagonists of scientific reasoning and research for several hundred years, were redirected by one of their own scholars, Abu Hamid Al-Ghazali, who lived from 1055 to 1111, to spend more time studying theology than science. Following that, the textbooks of this progressive Islamic school, with students from around the entire world, were transcribed into Latin by Albertus Magnus (1200 to 1280), and he and his student, Thomas Aquinas, continued with the study of science but now from a Christian perspective.

The Islamic example I have provided represents a very small segment of their religion, and it should be noted that this group is very likely extremist in practice. Those who live under and practice extremist values will suffer educational and cultural difficulty, not to mention a complete lack of understanding of science. This will never change for anyone in this category until they are able to properly redirect their thinking to the difference in theology and science with trust in both.

Resurgence in Flat Earth Beliefs

Flat Earth belief, religious and otherwise, has been making a comeback recently in almost every society and spreading rapidly through social media.[55]Conspiracy theories appeal to those who believe that they alone are in possession of simple truth, and they avoid complex statements because they do not trust or understand them as even being possible. Perhaps they also believe that they alone are in possession of data that the powerful wish to suppress. Sadly, less-educated people are among those who do not trust public institutions. They are extremely common in autocratic or theocratic societies, where they are either told or even assume that other authorities outside of their idealistic realm of influence are lying. Simplistically, it could even be called brain washing, alternate truth, fake news, and so on.

Skeptics of all beliefs and in all societies will often claim the horizon will always remain completely flat to the observer, regardless of how high they travel. Also, they believe that people who claim to see the Earth's curvature from a plane are lying. I refer you back to the introduction of this section as someone who has been there and seen the proof with my own eyes. Clearly, the science agrees with my own eyes.

Sadly, religious fundamentalists are not alone and those who have chosen with free will to believe that NASA is conspiring to provide false information will also be left behind in Darwin's world of survival of the species.

---

[55] Google this subject through this attached locator, and also look up further variations of this subject, https://www.tfes.org/

Stop, Think, Discuss

Think about the reasons all theology historically would want to create a story of origin to educate or tell people who mostly didn't read or write. What would be the purpose of such a story and why would it be created?

Question one: Do you think God wants us to believe and learn in science? Is God even involved in science, as it changes the theological thinking of major religions? Is science God's creation or man's curiosity?

Question two: Where does science become religion, and theology try to become science in all these units of information within this chapter?

Question three: What is more important to you in all this: the proofs of science or the cultural adhesion of theology with all the subjective thinking of ethics and morality?

Question four: Absolutely everything that happens has a precursor that makes it happen. Then the event happens. What made *Homo sapiens* want to know more about history, science, and structure with the overriding adhesion of love, artistry, justice, and theology?

Question five: What is your opinion of religion thinking it is science or science thinking it is religion? Do you think they overlap? If no, then explain.

Question six: Are there statements within this chapter that are beyond belief? Does this epistle of information exceed common sense?

CHAPTER 28

They Looked, Considered, and Provided Their Solutions

Anaximander of Miletus (610–546 BCE) viewed the Earth as flat, like the face of a cylinder, with a thickness one-third its diameter, like a pancake. For that reason, the Earth is suspended and remains in place because it is equidistant from all other things. It cannot fly off in any direction. He also postulated that the sun and the moon are hollow hoops around the Earth, filled with fire. Imagine the hoops of a barrel. The sun and moon's round, disk-like appearance are holes in the hoops, through which the fire can shine. The phases of the moon, as well as eclipses of the sun and the moon, are due to closing of the vent's holes, or vents in these rings of fire closing.

Along came Pythagoras (569 to c. 490 BCE) who promoted the idea of "cosmos", where bodies move in perfect circles. He was followed by Aristotle (384 to 322 BCE), a student of Plato, who in turn was a student of Socrates, who bought into the Pythagorean principle but stated that the Earth was the center of everything. Two hundred years later, about 220 BCE, a sun-centered model was proposed, and twenty years after that, the circumference of the Earth was proposed by Eratosthenes (276 to 196 BCE), who calculated the circumference of the Earth without leaving Egypt. Then Archimedes (287 to 212 BCE) discovered the formula to calculate the circumference and volume of a sphere, like the Earth. Whoops! The Earth is now a round sphere.

In 150 AD, Ptolemy writes about an Earth-centered model again, which is back to Aristotle and company, and Ptolemy's theory was the way it was for nearly 1,400 years. Remember, science became centered around Arabic and Islamic Bagdad until roughly 1200 AD, and then Thomas Aquinas and the Christian Church after 1200 AD, mostly because of the power of the Roman Christian empire. Yes, there were schools of thought that were opposing this Earth-centric idea, but they were soon shut down.

After Greece and into the Dark Ages

And then we enter the dark ages of the Western world and the scientific advancement of the other regions of this planet.[56]

400	Away from Greece and the Christian Roman Empire, others were happily studying the stars. In India, Hindu cosmological time cycles explained in the Surya Siddhanta, year 400 AD, gave the length of the year as 365.2563627 days. Only 1.4 seconds longer than the modern value of 365.256363004 days. This was the most accurate measurement for the length of the year anywhere, for more than a thousand years.1
476 to 550	Indian mathematician and astronomer Aryabhata (476 to 550 AD) identifies the force "gravity" to explain why objects do not fall when the earth rotates. He proposes an earth-centered solar system of gravitation and an elliptical model for the planets. The planets spin on their axes and follow elliptical orbits; the sun and the moon revolve around the earth in interacting orbits. He states that the planets and the moon reflect the light of the sun, and the Earth rotates on its axis, causing day and night. Somehow the "centered" earth rotates around the sun causing years.1

---

[56] The creation of this table was heavily influenced by material from https://en.wikipedia.org

628	628 AD, Indian mathematician and astronomer Brahmagupta recognizes gravity as a force of attraction. He describes the second law of Newton's law of universal gravitation, and Newton is not even a twinkle in his father's eye. He provides methods for calculations; the motions and locations of our five planets, their rising and setting, conjunctions, and of the solar and lunar eclipses.1
777	Then in 773 through 777 AD, the Indian Sanskrit works of Aryabhata and Brahmagupta are translated into Arabic, giving Arabic/Islamic astronomers an insight into Indian astronomy.1
850	By 850 AD, the first major Arabic/Islamic work of astronomy exists. It contains tables for movements of the sun, the moon, and the five planets known at the time. The work is significant as it introduced Ptolemaic concepts into Arabic/Islamic sciences. Remember, Ptolemy makes Earth stationary and at the center of the universe. This Ptolemy concept introduced into Arabic/Islamic science is the turning point for their astronomy. Prior to this, they only translated works of others learning from their knowledge. This new Arabic/Islamic work marked the beginning of new ways to study and experiment with calculations.1

**Eighth to
thirteenth Century**

In the eighth century, the Islamic city of Baghdad became the greatest learning center on the earth. It welcomed all religions, philosophies, and scientific study to its universities and openly allowed the use of its libraries. For five hundred years, the outpouring of original science from Baghdad was beyond anything the world had ever seen. To this day, the science is highly regarded. More than two thirds of the stars in the sky have Arabic names because they were discovered by Arabic astronomers. Arabic numerals were invented, Roman numerals were discarded. This era of great discovery declined significantly in the eleventh century when the scholar Hamid al-Ghazali wrote a series of papers questioning the philosophy of Plato and Aristotle and declared mathematics to be the philosophy of the devil. This precipitated a series of events that undermined scientific thinking because now Islamic theology became a compulsory study and sadly the entire Arabic/Islamic scientific movement in Bagdad collapsed.

850

Back to 850 AD and an Islamic book by al-Farghani; "*A compendium of the science of stars.*" The book primarily summarized Ptolemaic theories. However, it also corrected Ptolemy, based on works of earlier Arab/Islamic astronomers. The books were distributed throughout the Muslim world, and even translated into Latin.1

928 to 1031	In 928 AD, one of the first astrolabes is constructed by Islamic mathematician-astronomer Mohammad al-Fazari. Astrolabes are the most advanced scientific tools of their time. The precise measurement of the positions of stars and planets allows astronomers to compile the most detailed almanacs and star atlases to that time. By 1031 AD, Abu Said Sinjari, a contemporary of Abu Rayhan Biruni, suggested the possible movement of the Earth around the sun, i.e. heliocentric.[1]
1054	Chinese astronomers in 1054 AD record the sudden appearance of a bright star. Native American rock carvings also show the brilliant star close to the moon. This star is the Crab supernova exploding.[1]
1070	In 1070 AD, one of the most important works of the period was Al-Shuku ala Batlamyus ("Doubts on Ptolemy"). In this, the author summed up the inconsistencies of the Ptolemaic models. Many astronomers took up the challenge posed in this work, namely to develop alternate models that evaded such errors.[1]
1126	In 1126 AD, Islamic and Indian astronomical works (including Aryabhatiya and Brahma-Sphuta-Siddhanta) are translated into Latin in Córdoba, Spain, introducing European astronomers to Islamic and Indian astronomy.[1]

1150	Indian mathematician-astronomer Bhāskara II, in 1150 AD, calculates the longitudes and latitudes of the planets, lunar and solar eclipses, risings and settings, the moon's lunar crescent, and conjunctions of the planets with each other and with the fixed stars, and explains the three problems of diurnal rotation. He also calculates the planetary mean motion, ellipses, first visibilities of the planets, the lunar crescent, the seasons, and the length of the Earth's revolution around the sun to nine decimal places.

With the focus of science now a low priority among the Arabic/Islamic leadership, science was pushed to the side. Scholars from India and China continued working. The translation of the Arabic scientific texts into Latin and the Christian acceptance of science transformed the next wave of scientific leaders to independent thinkers within the Christian world.

Christianity Trades Places with Islam

St. Albertus Magnus, Christian Saint Albert the Great, (1193–1280) was a Dominican bishop and philosopher best known as a teacher of St. Thomas Aquinas and a proponent of the philosophy of Aristotle at the University of Paris. Magnus established the study of nature as a legitimate science in Christian tradition. He was the most prolific author of his era and the only scholar called "the Great" while alive.

He discovered the works of Aristotle, which had previously been translated from Greek and Arabic before 1245.

During his time in Paris, Magnus worked diligently on creating a book presenting the entire knowledge of the world known at his time. No other medieval scholar made commentaries on all the known works of Aristotle, and like this presentation, paraphrasing the originals and adding phrases to make one think, by expressing observations, speculations, and a form of experiments. Experiment for Albertus was a process of observing, describing, and classifying. Apparently, he was responding to a request to explain physics, as proposed by Aristotle. His undertaking was to make understandable to his society the major works of science that were unknown prior to this: natural science, logic, rhetoric, mathematics, astronomy, ethics, economics, politics, and metaphysics. He taught and worked on this encyclopedia for nearly twenty years, and one of his students was Thomas Aquinas.

Saint Thomas Aquinas (1225–1274) was an Italian Dominican friar, Catholic priest, and Doctor of the Church. As a philosopher, Aquinas carried significant influence within the Roman Catholic Church as a theologian, with a tradition of "scholasticism", which in Christian terminology means medieval university teachings on Aristotelian philosophy, with heavy emphasis on Church tradition and dogma.

Thomas Aquinas was the primary proponent of theology, which is based on facts and experience, called Thomism, wherein he argued that reason is found in God. Thomas embraced numerous ideas put forward by Aristotle, whom he called "the Philosopher", and he did his best to incorporate Aristotelian philosophy with principles of Christianity. These explanations on Scripture and on Aristotle form a significant part of his work, which became integral to the Catholic Church's theology in the twelfth century.

As a result, the Roman Catholic Church honors Thomas Aquinas as a saint and regards him as the model teacher for those studying for the priesthood. Thomas Aquinas is considered one of the Catholic Church's greatest theologians and philosophers of all time.

Age of Science

In 1350, (1304–1375), an Arabic/Islamic Syrian Astronomer, engineer and mathematician in Damascus, wrote *A Final Inquiry Concerning the Rectification of Planetary Theory* after calculating trigonometrically that the Earth was not the exact center of the universe. Ignoring Ptolemy because his eyes told him otherwise, al-Shatir came up with a planetary arrangement that has Earth almost in the center but not exactly so. One hundred years later, the Copernican model was generated using his principles.

In 1543 AD, Nicolaus Copernicus published his scientific findings containing his theory that the Earth travels around the sun, called the "heliocentric" model. By using Plato's perfect circulatory model of moving planets, however, he missed the true result, thus complicating his theory because those watching the heavens, and Mars specifically, came to understand that it did not visually comply with a circular orbit.

Incidentally, and as a side bar, Socrates taught Plato, who in turn taught Aristotle, who in turn mentored Alexander the Great. And as we read earlier, Aristotle became the scientific potentate of the Roman Catholic Church because of Saint Thomas Aquinas, who proposed that Aristotle was correct in his thinking.

The Roman Catholic Church, in the sixteenth century, was planted in the works of Aquinas and his fixed interpretation of Aristotle. New interpretations were impossible, and the new revolutionary movement of Protestants only fixed the Catholic Church in their position. Alternate interpretations of scripture and/or science were banned. The Church had become polarized and was about to come apart at the seams due to a lack of insight into the needs of some of its own clergy and parishioners.

With this timeline, we see when Christian theologians recognized the scientific significance of Albertus Magnus and Thomas Aquinas and allowed these two brilliant (theological) students to investigate outside the formidable restrictions of the Christian faith, which from 400 to about 1200 put the Western world into the "Dark Ages". Only by allowing Albertus Magnus and Thomas Aquinas freedom to learn from the Arabic/Islamic and Indian scientific precepts, learned in the schools in Baghdad, did the Western world come out of the Dark Ages from 1200 onwards.

Stop, Think, Discuss

Try to imagine the mental acumen of someone in 650 BCE looking at the sky and coming up with bands of fire contained within rings around a flat Earth venting out light in small holes, which were the sun and moon. Now ask yourself what you might have thought about in that time prior to knowledge and written text,

Also ask why it is important to think about creation in the form of the Earth, sun, and moon and everything else.

Question one: Do you think a timeline gives clarity to the process of discovery?

Question two: The great educational and scientific flip flop occurred when Islamic leaders dismissed their own scientific leaders to redirect study of more theology and Roman Catholic Christians took over leadership of scientific investigation after hundreds of years of fixed Christian, theologically finite, thinking. Is this coincidence? Does God have a hand in the changeover?

Question three: Aristotle was not a great physicist, especially when compared to Archimedes. Pythagoras and Aristotle were at odds about the creation of babies, and the Roman Catholic Church adopted the (modified) Pythagorean model of babies made by male sperm and nurtured by female wombs (the homunculus, as discussed earlier). How can the Roman Catholic Church justify adopting the science of Aristotle from Thomas Aquinas, yet deny his philosophy in the study of DNA?

Question four: Absolutely everything that happens has a precursor that makes it happen. Then the event happens. Why would most of the leadership of the Arabic/Islamic world accept that theology was to be the only basis of scientific study, thus terminating a 500-year hiatus of world leadership? Could you have determined, before this elemental change of direction, the cause of the event? Also, could the event have been prevented if the cause had been known beforehand?

Question five: How would you describe the timeline of events in your own words? What significant changes have been caused by all these events, and have these events changed the world for better or worse?

Question six: If you read children's fiction, Tinker Bell made the boys fly with pixie dust. What kind of dust do you see in this chapter?

CHAPTER 29

The Renaissance

The Renaissance was a primordial soup generating the advance of Western Civilization—arts, culture, theological change, medicine, philosophical thinking, and the beginning of the age of correct science, unhindered by the burden of theological dogma and the power of the Church.

When Nicolaus Copernicus published his paper in 1543, he had been showing it around to other scientists, afraid of what would happen if the very powerful Catholic Church had other ideas about this, declaring that the Earth was at the center. Besides, Aristotle said so, and we see from the timeline in the previous chapter, that Aristotle (mentor of Alexander the Great) was now a potentate of thinking scientifically within the Catholic Church. So, Copernicus waited until he was on his death bed before publishing his paper. You can't do much to those who are no longer with us. But not all theologians were detractors. Copernicus's work was used by Pope Gregory XIII to reform the calendar in 1582

Along came the great Dane, Tycho Brache (1546–1601), a wealthy nobleman who had benefited with a great education. Meticulous in observation and dogmatic in work ethic, he created a sun-centered model of our universe that met all criteria of his astrological observations. He built the finest naked-eye observation instrument of the period and maintained the most accurate measurements of planetary positions ever. At this point, it was perhaps possible to see as many as four thousand stars and objects in the night sky. Bache educated others in his astro-laboratory. It was not until 1608 that telescopes were invented.

One of his brilliant students, a German by birth, came to study under his tutelage. This student, Johannes *Kepler* (1571–1630), was brash and outspoken, and often a pain to deal with. The model that Brache had created was working perfectly well except for one feature. As anyone will tell you, if planets orbit the sun in perfectly circular orbits, they will change positions progressively forward every day as they go around the sun. However, as even the Babylonians discovered, Mars does not cooperate. Mars moves backward to the eye as it goes into certain positions of its orbit around the sun. This is used in astrology, for instance. It also is not feasible in the perfectly centric universe that Bache was modeling.

So, Bache gave his imperious student, Kepler, this problem to solve. It took several years for *Kepler* to resolve this dilemma, but he did, and behold he found that the orbits of the planets are not circular, they are elliptical. Using this model, all planets obeyed the observations of the eye perfectly! Kepler went on to publish his first and second laws of planetary motion in 1609, and his third law later.

Meanwhile in the land of tulips, an eyeglass maker named Hans Lippershey, who also had two lips, invented the telescope in 1608. Galileo Galilei (1564–1642) instantly made extensive use, both for personal gain and for study, of this pivotal moment in the history of astronomy. Galileo made improvements to this telescope with his first to be about 3x magnification. Later, he improved it with up to 30x magnification. With a Galilean telescope, the captain of a ship could see magnified images upright on the ocean. This was a profitable business, and it was sought after by trades people and ships' merchants everywhere. It was coined a spyglass. But it could also be used to look at the stars. In his time, he was one of only a few who could build telescopes good enough that the stars were observed with clarity. On August 25, 1609, he demonstrated one of his early telescopes, with a magnification of about eight or nine. He used his telescopes for more than trade and studied the night sky. He first published his telescopic astronomical observations in March 1610.

Opposition to Galileo's observations came from both religious and scientific quarters but was mostly stoked by religious political happenings. The Christian reawakening was in full progress, and strife within the Church was paramount. The Protestant revolution was causing the Roman Catholic Church to "lock down" principles of interpretation and no other insight than that of their masters was tolerated. Religious opposition to the sun-centered universe ("heliocentrism") arose from Aristotelian Bible interpretations as espoused by Albertus Magnus and Thomas Aquinas.

Perplexed in all of this, Galileo wrote a letter wherein he argued that heliocentrism was not contrary to biblical texts, and further, that the Bible was an authority on faith and morals, not on science. In today's world, this is an interesting point of view that is obviously recognized as reality. But in February 1616, an Inquisitorial Commission declared heliocentrism to be "foolish and absurd in philosophy, and formally heretical since it explicitly contradicts in many places the sense of Holy Scripture." On February 26, Galileo was ushered to the Cardinal Bellarmine residence, who demanded that he, "abandon completely" his opinion that the sun stands still at the center of the world and the Earth moves.

Cardinal *Bellarmine* had served as rector of the Collegio Romano in 1592, as provincial of the Neapolitan province of the Jesuits in 1594, and papal theologian in 1597. He was made a *Cardinal in 1599*. Galileo was to henceforth not to hold, teach, or defend his opinion in any way whatever, either orally or in writing. During the Roman Inquisition, Cardinal Bellarmine was made Cardinal Inquisitor where he served as one of the judges at the trial of Giordano Bruno in 1600 AD and concurred in the decision that condemned Bruno was to be burned at the stake as a heretic for holding the Copernican model of the universe and numerous other sins. So, this was a very stressful time for Galileo. Remember, the Reformation had been started in 1517 by Martin Luther and was successful in attacking the power of the Roman Catholic Church. The Catholic Church was trying to consolidate and regain power within its ability to do so. Burning at the stake and heresy were all too common as methods of forcing compliance.

For ten years, Galileo avoided this controversy. Then he renewed his project of writing his book on the subject when his friend, Cardinal Maffeo Barberini, was elected Pope Urban VIII in 1623. Barberini had opposed the condemnation of Galileo in 1616. Galileo's resulting book, *Dialogue Concerning the Two Chief World Systems*, was published in 1632 and received formal authorization from the Inquisition with papal permission.

Whether unknowingly or deliberately, Galileo used Simplicio, as a defendant character of the Aristotelian geocentric view. This character was interpreted to be the pope, his friend. In Italian literature, Simplicio was often caught in his own errors and often was observed as a fool. So, this comparison to the pope was not complementary and not received well by his now-previous friend, the pope. Indeed, although Galileo states in the preface of his book that the character is named after a famous Aristotelian philosopher (Simplicius in Latin, "Simplicio" in Italian), the name "Simplicio" in Italian also has the connotation of "simpleto". Unfortunately for his relationship with the pope, Galileo put the words of Urban VIII into the mouth of Simplicio.

It is agreed by most Christian historians that Galileo did not act in malice and felt unprepared by this reaction to his writings. Regardless, the pope was not amused and took the perceived criticism the same way Donald Trump would accept it. With malice.

Galileo offended the pope, his biggest and most authoritative supporter. He was ordered to Rome to defend his writings in September 1632. He took his time and arrived in February 1633. He was brought before Inquisitor Vincenzo Maculani to be charged. In those days, it would be frightening for sure. People were burned at the stake after all! All through his trial, Galileo maintained that since 1616 he had faithfully kept his promise to not hold any of the "Church condemned" opinions. Regardless, they wore him down, and he eventually admitted that a reader of his *Dialogue* could well have concluded that what he wrote was a defense of Copernicanism, obviously, clearly opposite to his true intention. His last lengthy interrogation in July 1633 ended with threatening him with torture if he did not tell the truth, and he steadfastly maintained his denial despite the threat. Heretic burn, baby, burn!

The sentence of the Inquisition was delivered on June 22. It was in three essential parts. First, he was convicted of the heresy "of having believed and held the doctrine—which is false and contrary to the sacred and divine Scriptures—that the sun is the center of the world and does not move from east to west and that the Earth moves and is not the center of the world" (Rose 2016). Second, it was said "that an opinion may be held and defended as probably after it has been declared and defined to be contrary to the Holy Scripture" (Rose 2016). Third, it stated "that consequently you have incurred all the censures and penalties imposed and promulgated in the sacred canons and other constitutions, general and particular, against such delinquents" (Rose 2016).[57] Galileo was found "vehemently suspect of heresy" for holding the opinion of "heliocentrism", that the sun is the center of the universe and the Earth

---

[57] These three quotes are from *The Protestant's Dilemma* by Devin Rose, published February 27, 2014.

is not, and that he held and defended this opinion as likely, even after it had been declared "contrary to Holy Scripture" (Rose 2016).

Galileo was sentenced to formal imprisonment at the pleasure of the Inquisition. The following day, this sentence was commuted to house arrest, obviously clemency from his previous friend the pope. He remained under house arrest for the rest of his life. His offending *Dialogue* was banned, and with malice of those not his friends and not announced at the trial, any further publication of any of his works was forever forbidden, including any he might write in the future. Well, that is one way to eliminate opposition when your opponent is politically connected, and you are not.

After time with the Archbishop of Siena, Galileo could return to his villa at Arcetri near Florence in 1634. There he spent the remainder of his life under house arrest. Additionally, Galileo was ordered to read the seven penitential psalms once a week for the next three years. However, his loving daughter and a sister in a Catholic order, Maria Celeste, relieved him of the burden after securing ecclesiastical permission to take it upon herself.

While Galileo was under house arrest, he dedicated his time to one of his finest works, *Two New Sciences*. In these works, he summarized what he had done some forty years earlier, now called kinematics and strength of materials. It was published in Holland to avoid the Catholic censor. Much later in time, this book received high praise from Albert Einstein, and strength of materials was a wonderful subject I studied in engineering classes.

Remember, Europe in this era of Galileo was in great turmoil. The Roman Catholic Church, which had gained status and power from Emperor Constantine in 315, had managed to become all-powerful over the previous 1,400 years. Now it was under siege by the mainstream population, who were fed up with their all-powerful and controlling ways, which even included life and death. The bishops were often the sons of the potentates who were governing the region, and who often never set foot in their own diocese. These same bishops controlled the succession of who would govern or reign. They were given large stipends by the Church for their own purposes. Celibacy was the supposed pious lifestyle, but many of these Church leaders and aristocrats worked around this by being either lecherous predators or engaging "live-in" housekeepers.

Martin Luther in 1517, along with the Gutenberg press, had declared war upon the establishment by starting a Protestant movement that quickly enveloped the masses who were revolting. The bishops, barons, aristocrats, and all-powerful—at least in their own minds—were not going to let their lofty status and lifestyle go. Besides, there was money involved for these potentates, lots of money, and very little of it earned through their own ability or effort. Mostly funds came from taxation of the poor. The Church had translated the Bible into Latin, a language only understood by wealthy aristocrats. Then they often used their own versions of interpretation to separate funds from the lower classes; versions that were nearly nonexistent in the written word of the Bible or which were greatly misrepresented. The Protestant movement used common language to interpret the Bible, a biblical interpretation that was rejected by the well-educated elite of society.

> *He that has ever so little examined the citations of writers cannot doubt how little credit the quotations deserve when the originals are wanting.*

> Martin Luther[58]

---

[58] Martin Luther, O.S.A. was a German professor of theology, composer, priest, monk, and a seminal figure in the Protestant Reformation. en.wikipedia.org

Stop, Think, Discuss

All these actions are just another example in human history of how some brilliant *Homo sapiens* have made futuristic discoveries in science, and then others, in this case the powerful potentates of the Church, with less than altruistic ideas, have used or disputed the science for their own warped desires of power, wealth, or status. This would seem to be the Ying and Yang of discovery, or the third law of Newton: To every action there is an equal and opposite reaction.

How much of this discussion do you see as a power play by authorities losing control and trying to regain it versus true objection to the science being presented?

Think of science that has been wrongfully modified by theology. Now think about theology not having influence in science and how that has gone wrong. Think eugenics, and the ethics, morals, and commonsense that were wanting in that field.

Question one: In these statements about geocentricism and heliocentrism, is there divine intention of God and where is the hand of man? Is God even involved? Is this whole history God's creation or man's? Or a combination of both?

Question two: What is the *ontology* of this history? That is, what is the actual information, free from our personal bias, of these statements? And what is the *epistemology*? That is, what is the knowledge, as learned by us, including our preconceived notions, of these statements? Do opinions reflect the author's intention or are they free from opinion?

Question three: Go back to Aristotle. Then think of the progression forward in the science of the stars. What appears to come first in this? Was it an idea created by God or by man?

Question four: Absolutely everything that happens has a precursor that makes it happen. Then the event happens. Could you have determined, before the event, the cause of the event? Also, could the event have been prevented if the cause had been known beforehand? What happened in the beginning to cause the superior thinking of the philosophers of ancient Greece? And what about the Arabic/Islamic schools in Baghdad and their subsequent demise after 500 years, giving rise to Christian scientific thought in the modern age? Could this involve God or just be coincidental? If coincidental, what are the statistical odds of that happening over all the ages in the way it has happened?

Question five: Provide your testimony of these events in your own words.

Question six: Do you think Harry Potter has arrived on the scene in the viewpoint written in this chapter? Has creative thinking taken over the dialog on this issue?

CHAPTER 30

Newton and the Age of Breakthrough Discovery

The telescope changed everything. Galileo demonstrated how it could be used to observe the stars before he was admonished by the all-powerful Church for his work. By 1656, Christian Huygens discovered Saturn's rings and Titan, the fourth moon of Saturn. Then he noted markings on Mars in 1659. Shortly afterwards, another observer, Cassini, discovered polar ice caps on Mars.

Then, out of the English countryside, galloping into the history of our species, came "rootin' tootin'" Newton! He was born in 1642, died in 1672, and is considered by most scientists to this day to be the most brilliant scholar of the modern era.

His counterpart of the ancient era is Archimedes, who was a Greek mathematician, physicist, engineer, inventor, and astronomer (287 BCE–212 BCE). Archimedes, who lived in Italy, was the discoverer of many things, including the mathematical constant for circles "pi" or 22 / 7 = 3.14. Note: History also indicates that the Babylonians and Egyptians also discovered the diameter of a circle. Archimedes was then able to produce formulas for determining the area, volume, and circumference of circles, spheres, and globes. Archimedes published a book called *The Sand Reckoner*, wherein he counted the number of grains of sand in the universe. He did this in his work to perhaps counter the Jewish Bible's question, "Who can number the sand of the sea, and the drops of rain, and the days of eternity?" (Eccl 1:2 NIV). He was murdered by the Romans in the battle of Syracuse in 413 BCE.

Back to the future. Isaac Newton had a weak persona. He was not someone who could withstand the withering criticism of detractors or powerful potentates who would consider his ideas as disparaging to their own. His ideas would not survive a regime like that which pursued Galileo, for instance. So, it was for many futuristic thinkers of our history. The minute they represented a challenge to the authority of those in command, they would be put down, and the world would miss the stepping stones of creative discovery into the future.

It wasn't until after a visit from his good friend, Edmond Halley, as in Halley's Comet, that Newton considered publishing his breakthrough in science. Halley came to him in 1680 to discuss his own ideas of why the comet was seen going to the sun and thereafter away from it. Newton did the math and calculated the force to acceleration and insisted that the strength of the force varied as to the inverse square of the distance, as correct. This so impressed Halley that he went on to calculate twenty-four other comets in the solar system and insisted that Newton publish his works.

Newton was accomplished in numerous principles, including being called the father of calculus. He developed his law of gravitation in 1666 at the age of twenty-three. He developed his three principles, often called Newton's three laws:

1. An object at rest will stay in that state until acted upon by an external force.
2. A body in motion will not change speed or direction until acted upon by a force external to it. Force equals mass times acceleration. .
3. For every action there is an equal and opposite reaction. He used the example of a man swinging a ball, in a circle held by a rope, to demonstrate centripetal force of gravity between celestial bodies.

Newton studied optics and invented the reflecting telescope, making this Dutch invention a new and improved instrument. In this, he studied the prism and the spectrum of white light. He taught physics and astronomy, and his book set out the principles of gravity and celestial mechanics. He formalized the physics of the speed of sound and explained why the Earth is a squashed sphere.

Like a lot of gifted thinkers, Newton also spent a lot of his time in other fields, including historic biblical chronology as well as alchemy.

The discovery of the universe was now quickly unraveling mystery after mystery. From the mid-1700s to 1916, the significant following (there are many more) astronomical items were discovered.

+ Uranus is discovered by Herschel in 1781
+ Piazzi discovers the first asteroid, Ceres, in 1801
+ Discovery of the Doppler Effect in 1842
+ Johann Galileo observes and discovers Neptune in 1846

- The Great Red Spot on Jupiter becomes prominent in 1878
- Henrietta Swan Leavitt discovers Cepheid variables in 1908
- Albert Einstein introduces his general theory of relativity 1916

Taking Credit for Discoveries

We, the lesser educated of the species, call laws discovered by various scientists by the name of the scientist discovering that existing rule of force. Newton's three laws, Bernoulli's principle, Gauss unit, Boyle's law, Einstein's theory of relativity, and so on. This is representative of our narcissistic human ego taking credit for discovering what already exists, which is revealed to us through some form of mysterious human endeavor, ignoring the fact that it is preexistent to our history within the universe. We are not unlike the ancients who saw the Earth as flat and stationary, with everything in the night sky revolving around it. But like them, we are centrist. We create the illusion that everything is about us.

This is where there is, or should be, a God effect. We *Homo sapiens* did not create these scientific laws; they are finite rules of order that made possible the whole creation of the universe, starting with the big bang. So how did these rules come to be? As expressed earlier, this is a deist point of view.

On the other hand, we (if you are Christian or Jewish) call the Ten Commandments "laws"'", when they are merely suggestions for living a great life, which helped *Homo sapiens* grow into the community of beings we presently are. This is truly a confused state of mind between scientific thought and theology. And perhaps ego. Consider, when is a scientific law breakable? Not until someone finds alternate physics that enlarge the concept of the law; until then, it is fixed, finite, and unbreakable—like the law of gravity, for instance. Alternatively, the law of man is completely discretionary as to whether you can break it or not, and it is not possible to continuously obey the Ten Commandments. Thus, the reason for forgiveness. Further the Ten Commandments can be used to postulate an unfailing God who is able to keep his own commandments. They are also a guidepost to humans as to how to live life and provide warnings of when they are going wrong. Yet for others, they are a suggested way to try to constantly live the laws of an unfailing God by demonstrating good works, which only leads to guilt when the inevitable failure to keep them occurs. There is much philosophical discussion around this interpretation, especially the third use of the law.

Stop, Think, Discuss

Scientists, mathematicians, physicists, and so on, are always considered as being the smartest *Homo sapiens* in the world. Have you ever considered whether humans in other fields could attain this elevated status? Could it be possible for a teacher or philosopher or theologian to be considered one of the smartest?

Question one: Reading about scientists and other brilliant philosophical scholars, many seem to have weak personalities. How would they fare in societies that were alpha-male dominant? How would society fare with the same situation? Would there be historical progress in society or regress and is this predetermined by any God influence?

Question two: One of the statements about Newton was his interest in alchemy. Is this the only area of science that he failed at? Did he find the Midas touch?

Question three: Where does there appear to be a God in this information? Was this scientific progress created by God or by man? What is your opinion unbiased by previous teachings?

Question four: Absolutely everything that happens has a precursor that makes it happen. Then the event happens. Could you have determined, before the event, the cause of the event? Also, could the event have been prevented if the cause had been known beforehand? Do you think Newton did all his research without bias or information from earlier scientists looking at the same information?

Question five: How would you describe discovery of scientific fact in your own words?

Question six: Are there issues in these opinions with which you disagree? Explain.

CHAPTER 31

Hubble and the Movement of Galaxies

When the big bang theory was discussed and generally accepted by all, except some fundamentalist theologians, the movement of the constellations and the planets within was not understood in its entirety. The movement of these bodies was misunderstood to the degree that it was not until 1929 that Edwin Hubble[59] discovered that galaxies generally move away from each other faster and faster, depending upon the distance between them, increasing in speed as they move farther away. He used the "redshift" method of calculation. Redshift is using a prism and measuring the red spectrum of light from the galaxy and noting how it shifts back and forth on a level plane over time, using a telescope.

Hubble worked at Mount Wilson and proved that other galaxies existed outside of the Milky Way. He took photos using the observatory's Hooker telescope and compared varying degrees of luminosity among Cepheid variable stars.

> A Cepheid variable is a type of star that pulsates radially, varying in both diameter and temperature and producing changes in brightness with a well-defined stable period and amplitude.[60] (Wikipedia 2018) Classical Cepheids undergo pulsations with very regular periods on the order of days to months. Classical Cepheids are 4–20 times more massive than the sun, and up to one hundred thousand times more luminous. These Cepheids are yellow bright giants and supergiants and their radii change by millions of kilometers during a pulsation cycle.

---

[59] Edwin Powell Hubble was an American astronomer. He played a crucial role in establishing the fields of extragalactic astronomy and observational cosmology and is regarded as one of the most important astronomers of all time. en.wikipedia.org

[60] Wikipedia 2018

Classical Cepheids are used to determine distances to galaxies and are a means by which the Hubble constant can be established. Classical Cepheids have also been used to clarify many characteristics of our galaxy, such as the sun's height above the galactic plane and the galaxy's local spiral structure. (Wikipedia 2019)

The Hubble Constant is the unit of measurement used to describe the expansion of the universe. The cosmos has been getting bigger since the big bang kick-started the growth about 13.8 billion years ago. The universe, in fact, is getting faster in its acceleration as it gets bigger.[61] (*Encyclopedia Britannica*) Prior to this, the Milky Way's size was unknown. Hubble, through research, estimated that the Andromeda Nebula was almost nine hundred thousand light years away from our Milky Way. Therefore, Andromeda was its own galaxy. It was understood to be a spiral prior to that. The Andromeda Nebula was later discovered to be even much farther away, nearly 2.48 million light years, in fact. It was later renamed the Andromeda Galaxy.[62]

So, we are faced with the knowledge that some constellations are moving away from each other, and some at the speed of light. We know from measuring gravitational waves that they do collide with each other every several billion years or so, but not often. For instance, our Milky Way is going to collide with Andromeda about a billion years from now.

Why are some of the constellations moving away faster than the speed of light? This question has been resolved by explanation that, when we see the light from these constellations, it took several light years for that light to arrive. By the time we see the light, the constellation has moved several more light years away.

---

[61] Encyclopedia Britannica

[62] Much of this can be reviewed in; https://www.space.com/25179-hubble-constant.html. What Is The Hubble Constant? By Elizabeth Howell August 28, 2018 Science & Astronomy

Stop, Think, Discuss

Where are the stars going? If we return several billion light years from now, will the night sky only reflect the stars within the Milky Way? Is the universe infinite? Does infinite mean circular in nature?

Well, grasshopper, there is an opinion on these questions these days. In the next part, part five, we will discuss science of the quanta and within it there is a theory of quantum loop gravity. We have come a long way from the ideas of Anaximander of Miletus in 550 BCE.

Question one: In these statements about Hubble and movement of the galaxies, what is the divine intention of God? Is God even involved? Is this God's creation or man's?

Question two: What is the *ontology* of these statements? That is, what is the actual information, free from our personal bias, of these statements? And what is the *epistemology*? That is, what is the knowledge, as learned by us, including our preconceived notions, of these statements? Will the universe expand at the speed of light forever?

Question three: Until Hubble, the movement of galaxies was not known and was even believed to be random. So, what appears to come first in this? Was this an idea or action created by God, or by man, or coincidence?

Question four: Absolutely everything that happens has a precursor that makes it happen. Then the event happens. Could you have determined, before the event, the cause of the event? Also, could the event have been prevented if the cause had been known beforehand? The big bang is a burst of tremendous energy from something completely unknown. Is all the movement of everything after the big bang just the dissipation of this energy and thus a return back to the plasma state of our origin? Entropy? Explain your thoughts.

Question five: Describe this scientific discovery in your own words. What are the implications of this discovery?

Question six: Do you have a different opinion of this chapter and what is it? Explain.

CHAPTER 32
Theory of Relativity, Times Two

To this point in our human history, science is everything you can observe with the eye. Proof is what you can see, touch, taste, or feel. Enter Einstein, spotlight stage right, pun intended. Jewish in culture and religion, born in 1878 and died in 1955, Einstein was one of two pillars of modern physics—the other being Max Planck, discussed earlier.

Einstein revolutionized the way we look at the world. He received the Nobel Prize in 1921 as a result of his work in 1905 on the photoelectric effect of light particles. Until then, light was unknown to be anything other than possibly a frequency, but he proposed that light was composed of discrete particles. He predicted that shining light on a conducting metal surface would create an electric current. This was later proven in the laboratory and the result is the photoelectric cell.

Niels Bohr, a Danish scientist and contemporary of Einstein, developed a new model of the atom and discovered that many of Einstein's quantum ideas had to be incorporated into this revolutionary new model in the first decades of the century.

Einstein proved Newton's work correct in most of his discoveries; however, he also corrected Newton's theory of the effect of gravity on very large and very distant objects. Einstein came up with the correction that gravity is two theories—special relativity and general relativity.

Newton's rules work perfectly well for engineering and physical elements here upon Earth. But Einstein discovered that in space, not so much. Gravity bends and is not a straight line like the apple dropping. Light bends because it is made up of particles, which are affected by gravity. Time is variable, not constant, again because of gravity. Now, how about them apples!

General relativity says that large objects cause space to bend in the same way that putting a tennis ball on a thin sheet would cause the sheet to bend. The larger the object, the more gravity and space bend.

Special relativity says it is impossible to determine if you're moving unless you can look at another object.

Since gravitational pull affects time, and the closer to the center of a mass, the slower you go, then the less you age and are therefore younger. Time in the relationship with the Earth produces the fact that the core of the Earth is 2.5 years younger than the surface, since the big bang. This is one part of a formula called time dilation.

Putting this into a more simplistic perspective, it means that if your twin goes to the top of a mountain to live when you were born, and you stay at sea level, when you both reach old age, your twin on the top of the mountain will be older than you! This is all because of Einstein's discovery that gravity slows down time. The closer you are to a large mass, like the center of the earth, the slower time goes; thus, a person at the top a of a mountain ages faster than the person at sea level.

Part two of time dilation is concerned with speed. The faster you go, the slower time goes. This is not relevant until you go extremely fast, such as in a close fraction of the speed of light, which is 300,000 km per second. It is only when you approach these high speeds that time slows down quickly and will approach zero.

When is time not time? When it is spacetime, because time can speed up or slow down, depending upon gravity and speed, or motion. Do you remember how we discussed how to measure time by the rotation of the sun, which is something that is used in navigation? Now that science is correcting those values, albeit in a different context, which is not obviating the original value used in everyday practice.

Before Einstein, time and space and mass and energy were separate. But by bringing these then heretofore unrelated elements together, first in the concept of spacetime and immediately thereafter in the equation, Einstein completed his theory of special relativity. It took a spectacular mind to conceive special relativity; it is perhaps one of the least intuitive theories ever conceived in the history of science, yet it is central to physics.

With those three theories, Einstein changed the way we think about space, time, and matter, and the lesser thinkers in our midst reverted to sucking a thumb like an infant.

Einstein was also a human with a distinct and loveable sense of humor, having made statements such as these Einstein quotes.

We all know that light travels faster than sound. That's why certain people appear bright until you hear them speak.

Only two things are infinite, the universe and human stupidity, and I'm not sure about the former.

You never truly understand something until you can explain it to your grandmother.

Albert Einstein

But It Is Not Just Einstein

Discoveries abound. Nothing is finite. Everything is changing, and quickly.[63] In 1927, Jan Hendrik Oort proved that Sagittarius is the center of the Milky Way, our galaxy. Then, in 1930, Pluto is discovered by Clyde Tombaugh—although the debate is still open as to whether it is truly a planet or not. By 1931, Karl Jansky had discovered the presence of cosmic radio waves, and shortly after, in 1937, Grote Reber built the first radio telescope.

Following World War II, space gleaned the attention of the worlds' political powers and Russia was the first to launch manmade objects into orbit in outer space followed shortly by the United States of America.

In 1957, Sputnik was launched by the Russians as the first manmade object to orbit the Earth, and not to be outdone, the United States quickly launched their first orbiter in 1958, called Explorer One. The race was on. To capture prowess and the attention of the rest of the world, Russia launched Yuri Gagarin into space in 1961 followed by John Glenn from the USA in 1962. Fixated in the task of outdoing one another for the attention of the rest of the world, Russia sent a vehicle that soft landed on the moon in 1966 with the Luna 9, and the USA sent Surveyor 1. Then came the crux du force. Launched with full television coverage from the launch pad

---

[63] After reading the following list of conquests all found in https://en.wikipedia.org, read another book, this time by an American, Neil DeGrasse Tyson: *Welcome to the Universe*, 2016, published by Princeton Press. It is a brilliant read by an unpretentious author.

in Florida, Apollo 11 landed on the moon with Glen Armstrong and Chuck Aldrin walking on its surface in 1969

In 1974, another astounding discovery was made in Ethiopia. Our oldest ancestor, "Lucy", a hominid who lived more than three million years ago, is unearthed. While looking into the future, we discover our past and can scientifically prove she is "us"!

Science and discovery accelerated from 1976 onwards. Now we are using scientific skills to interpret our place in this universe, which will either supplement or supplant our theological thinking of the past. The US Viking probes land on Mars in 1976. Then in 1977, using an airborne observatory called Kuiper, three scientists—James Elliot, Edward Dunham, and Jessica Mink—discovered the rings of Uranus. The telescope invented by the Dutch in 1608 and used by Galileo to the discomfort of theologically trained interpreters of science has moved a very long way from its origin.

In 1978, the discovery of Charon, the moon of Pluto, by James Christy and Robert Harrington, using a ground-based telescope gave more insight into our place in this universe. Having been launched in 1977, the twin explorers US Voyager 1 and Voyager 2 started sending data back to earth in 1978. They carry cameras, magnetometers, and other instruments to identify the structure and form of our surrounding planets and their moons: Saturn and its rings from Voyager 1, and Uranus from Voyager 2. They continue to move into interstellar space, each on its own path, continuously sending data back for us to disseminate.

Discovery also has a price. In 1986, the Space Shuttle *Challenger* exploded shortly after takeoff, killing all astronauts onboard. Similarly, in 2003, a takeoff problem caused a reentry failure of space shuttle *Columbia* once again with the loss of all onboard.

Looking for more and more discoveries and knowing that our atmosphere has clogged our view of the sky with particulate, the Hubble Space Telescope was put into low orbit from space shuttle *Discovery* in 1990. It is producing discoveries to this day and will continue beyond our time.

While the evidence of discovery of orbiting planets is now irrefutable The Vatican under Pope John Paul II in 1992 announced that the Catholic Church erred in condemning Galileo's work, which used the telescope to prove that the science of Copernicus was valid. Remember, Copernicus published his testament on his death bed for fear of being called a heretic and being burned at the stake for stating the

Earth and planets circle the sun, and 359 years later, Galileo was convicted of heresy for stating the correctness of Copernicus' theory as observed by his telescope.

Space Shuttle *Endeavor* was first used as a shuttle in 1992. It made 25 missions to outer space, and in 2000, while in orbit, made a detailed, global map of Earth obviously proving that the earth is clearly not flat. The *Endeavor* was retired in 2011.

In 2001, the NEAR (Near Earth Asteroid Rendezvous) spacecraft reached earth's nearest asteroid, Eros, and landed upon it with tools to detect its origin and likeness to asteroids landing upon earth.

By 2011, a land-based rover was operating on Mars and discovered evidence for water based on the geography of the planet in some locations, giving rise to much speculation and cause for calls to continue exploring our understanding of the solar system.

Stop, Think, Discuss

Science has rapidly eclipsed the old historical values of fixed thinking, or scientific theology, and is moving forward faster and faster every day. Theology in the pure religious sense is comparably stationary in explanation of itself, being mostly fundamental in its dogma.

How would you change the dialog of theology to keep up with science? Should theology keep up with science, and why or why not?

Does theology have a place in this discussion about physical science? Think back to another chapter about the science of eugenics and how theology might have caused a different outcome of that science if it had had more influence?

Question one: What in this chapter is the divine intention of God and what is the hand of man? Do you think God is even involved? Is this science God's creation or man's?

Question two: Is the science in this chapter free from our personal bias or that of the scientists discovering it? What is the science, as learned by us, that comes from preconceived ideas, or was it pre-directed by man? Was any of the science in this chapter invented by man?

Question three: What appears to come first in this? Was science a concept created by God, or by man?

Question four: Absolutely everything that happens has a precursor that makes it happen. Then the event happens. Could you have thought about space and time and gravity as a combined and interrelated topic in your head, like Einstein and others?

Question five: Describe this chapter in your own words, with theology and science combined in your point of view.

Question six: Are there issues in these statements with which you disagree? Has Harry Potter— creative thinking—taken over the dialog in this chapter?

CHAPTER 33

Mysteries of the Universe

Approximately ninety-five percent of the mass and energy composition of the universe is from material that science knows and cannot see. While it is mathematically identified using formulas of gravitational relationships, not one particle of it is observed. It is known as dark matter and dark energy. It does not emit light or energy.

Astrophysicists understand more about *how* dark matter and dark energy affect the universe than they know about *what* it is.

"Motions of the stars tell you how much matter there is," Pieter van Dokkum, a researcher at Yale University, said in a statement. "The stars don't care what form the matter is, they just tell you that it's there." Van Dokkum led a team that identified the galaxy Dragonfly 44, which is composed almost entirely of dark matter.

> The familiar material of the universe, known as baryonic matter, is composed of protons, neutrons, and electrons. Dark matter may be made of baryonic or non-baryonic matter. (Redd 2019)

The *Encyclopedia Britannica* describes dark matter as follows:

> As "dark matter," baryonic dark matter is undetectable by its emitted radiation, but its presence can be inferred from gravitational effects on visible matter. This form of dark matter is composed of "baryons", heavy subatomic particles such as protons and neutrons and combinations of these, including non-emitting ordinary atoms. (*Encyclopedia Britannica*)

Other candidates for dark matter include dim brown dwarfs, white dwarfs and neutrino stars. Perhaps super massive black holes could also be part of the equation? However, this unseen matter would require a more commanding role in the structure of the universe than has been observed by scientists to make up the missing mass. There are other suggestions that dark matter is more exotic.

Is it time for science fiction yet? Let's head in that direction for a minute, because there is discussion about particles that have never been seen.

Now we enter the minds of those in this field who have preconceived ideas of what dark matter could possibly be—such as another theory, which states that dark matter is composed of non-baryonic matter, with the most likely candidate being weakly interacting massive particles (WIMPS). WIMPS have ten to a hundred times the mass of a proton, so they qualify in the mass department, but the weak interactions with matter we can see make WIMPS hard to detect. Neutralinos, massive hypothetical particles heavier and slower than neutrinos, are also candidates, but like in science fiction, they have yet to be spotted.

Hmmm. Is any part of this going to end up in Marvel Comics?

When you get there, there isn't any there, there.

Zen Proverb

Yet another candidate are sterile neutrinos. Neutrinos are particles that don't make up regular matter. Neutrinos flow like a river from the sun, but rarely interact with normal matter. So, they pass through the Earth and us, and we don't even know it. There are three known types of neutrinos and one theoretical; known are the electron neutrino, muon neutrino, and tau neutrino, and the theoretical fourth is the sterile neutrino, which is suggested as possible dark matter. The story is that a sterile neutrino would only interact with regular matter through gravity.

According to a statement by the Gran Sasso National Laboratory in Italy (LNGS):

> Several astronomical measurements have corroborated the existence of dark matter, leading to a world-wide effort to observe directly dark matter particle interactions with ordinary matter in extremely sensitive detectors, which would confirm its existence and shed light on its properties. However, these interactions are so feeble that they have escaped direct detection up to this point, forcing scientists to build detectors that are more and more sensitive. (Gran Sasso National Laboratory in Italy; LNGS)

Still another stated possibility exists regarding dark matter science! All the laws of gravity that have been used to date need to be revised.[64]Could there be another possibility? Who knows?

Let's think about this for a while. If scientists can't see dark matter, how do they know it exists?

The calculation of the mass of large objects in space is made using mathematics of gravity in centripetal force by studying their motion. When the spiral galaxies were examined in around the 1950s, it was expected that the material in the center of the galaxy would move faster than the material around the outside. But what they discovered was that the stars moved at the same velocity in both the center and the outer edges. The only conclusion is that there is more mass in the galaxy than could be seen with the telescope. Further, gas in the galaxies also suggested it was necessary for more mass than could be seen. We now know that whole clusters of galaxies could not stay together if the mass within was what we can see with the telescope.

Through his genius, Albert Einstein showed that massive objects in the universe, through gravity, bend photons (light). Knowing this, these light-bending features can be used like a lens. By studying how light is bent by gravity, which surrounds objects that are not seen (dark matter), maps can be made of this dark matter. These maps indicate ninety-five percent of the universe is not seen.

Closer to home, in 2014, NASA's Fermi Gamma-ray Space Telescope made maps of the heart of the Milky Way in gamma-ray light, revealing an excess of gamma-ray emissions extending from its core. Dan Hooper, an astrophysicist at Fernmilab in Illinois, stated, "The signals we find cannot be explained by currently proposed alternatives and is in close agreement with the predictions of very simple dark matter models" (Hooper).

> [9] *"So I say to you: Ask and it will be given to you; seek and you will find; knock and the door will be opened to you. 10 For everyone who asks receives; the one who seeks finds; and to the one who knocks, the door will be opened."*
>
> Luke 11:9–10 NIV

---

[64] Information from an article in https://www.space.com/ 20930-dark-matter.html

Dark Matter Versus Dark Energy

I have been using dark matter and dark energy synonymously as ninety-five percent of the matter of the universe. Now we will break down dark matter and dark energy into their separate components. Although dark matter makes up most of the mass of the universe, it only makes up about a quarter of the composition. The universe is overwhelmingly dominated by dark energy. Dark energy makes up nearly three-fourths of the universe. The mystery of this is that science does not have a clue as to what it is or how it reacts or operates.

In 1929, American astronomer Edwin Hubble studied exploding stars known as supernovae and determined that the universe is expanding and has been expanding since the big bang. This was unknown prior to this. It was thought that everything was moving randomly about like feathers in the wind. We have already been told that the Andromeda Galaxy and our Milky Way are going to collide in about a billion years.

The next obvious question is how fast is the universe expanding? Basic human thought would lend itself to the idea that this expansion would eventually stop, so when, and how much is this expansion slowing right now? Guess what? Studies of distant supernovae indicate that the universe is expanding faster today than previously. The universe is accelerating in its expansion. Hold on, folks, this could be a real attention-getter!

Once again, to calculate this phenomenon, in the 1990s, two independent teams of astrophysicists turned their eyes to distant supernovae to calculate the possible deceleration. Surprise, surprise, they found that the expansion was still speeding up! This would mean that something exists in the universe to cause energy to overcome the effects of gravity. This energy was named dark energy.

This meant back to the blackboard to calculate the amount of energy necessary to overcome gravity. This calculation determined that dark energy required to cause this expansion must be sixty-eight percent of the universe. Dark matter makes up another twenty-seven percent. Do the simple math, which leaves the "normal" matter that we are familiar with, which only makes up less than four percent of the cosmos around us.

That gives us four percent of the known or visible universe as we might see it through telescopes someday. But even with the Hubble telescope in space, all the stars, planets, and galaxies that can be seen today make up proximately 2.5 percent of the 4 percent of the possible visible universe as we presently know it. The other ninety-six percent is made of stuff we can't see, can't detect, can't even comprehend.

The good news is that we are just starting to understand our position within this enormous cosmos. The bad news is that we are insignificant and certainly not the focal point of everything that goes on inside the universe. Too bad. Now I have put a pin in the ego of the 7.3 billion *Homo sapiens* who inhabit this little piece of sand within this great universal myriad of rock and fusion.

The cause is dark energy, and the effect is the spreading universe. This only gives limited information. The remaining questions are unanswered. Recent observations have indicated that dark energy has been consistent since the big bang, which gives partial insight into some of its properties.

We re-enter the minds of the scientists. One explanation for dark energy is that the universe is filled with a changing energy field, known as quintessence,[65] "a fifth substance in addition to the four elements, thought to compose the heavenly bodies and to be latent in all things" (Oxford English Dictionary). Another is that scientists do not correctly understand how gravity works. The leading theory of deep think-ing considers dark energy to be a property of space. Albert Einstein was the first to imagine that space was not empty. He also knew that more space (which is not empty) could emerge. In Einstein's theory of general relativity, he included a cosmo-logical constant, which accounted for a universe that was stationary, which scientists thought existed at that time. After Hubble announced his discovery of an expanding universe, Einstein called his constant his "biggest blunder".

But Einstein's blunder (a constant like the constant "pi" in a circle) could be the best explanation for dark energy. Predicting that empty space can have its own energy, the constant indicates that as more space emerges, more energy would be added to the universe, increasing its expansion. Read the previous sentence again; it is significant.

Now we are getting into the next volume on quantum science.

---

[65] One possible solution for dark energy is that the universe is filled with a changing energy field, known as "quin-tessence". https://www.space.com. "What is dark energy?"

Although Einstein's cosmological constant complements observations, scientists still don't know why; it just does.

NASA is currently investigating this issue as part of its Alpha Magnetic Spectrometer (AMS-02) and Fermi Gamma-ray Space Telescope missions.

Black Holes

Black holes are the next mystery of space. A black hole is a place in space where gravity pulls so much that even light (photons) cannot get out, just the opposite of the Hubble effect of the expanding universe. This black-hole gravity is so strong because matter has been squeezed into a very tiny space. This can happen when a star is dying.

Our sun will never turn into a black hole. Only stars with very large masses can become black holes. Our sun is not massive enough to become a black hole. When it does lose its energy, it will become a white dwarf. Even if our sun were a black hole, it would still be the same mass, just much smaller. The tiny black-hole sun would have the same gravity as the current sun. Earth and the other planets would orbit the black hole as they orbit the sun now.

Black holes are not space-roving star and planet eaters. The Earth will not fall into a black hole because there isn't a black hole near enough to the Milky Way to make that possible.

We can't see black holes because they retain photons. No photons means no light, so we can't see black holes. They are invisible. However, telescopes with special tools can find them. These tools can see how stars in proximity to black holes act differently than other stars.

Scientists think the smallest black holes formed when the universe began with the big bang, 13.8 billion years ago.

Very large or stellar black holes are made when the center of a very big star collapses upon itself, which causes a supernova—an exploding star that blasts part of the star into space before becoming a black hole. Scientists currently believe that super massive black holes were created at the same time as the galaxies within which they reside.

Albert Einstein first predicted black holes in 1916 with his general theory of relativity. The term "black hole" was coined in 1967 by American astronomer John Wheeler, and the first one was discovered in 1971.

The recent occupant of the University of Cambridge seat previously occupied by Newton, Stephen Hawking, posted a paper online prior to his death. He is one of the creators of the modern black-hole theory, which eliminates the idea of an event horizon, which is the invisible boundary believed to surround every black hole. Beyond this horizon nothing, not even light, can escape.

Taking its place, Hawking suggests an "apparent horizon,"[66] which can only hold the energy and matter temporarily before eventually releasing it. Even then, it will be released as a garbled quantity. Simply put and using the most recognizable equation of all time (from Albert Einstein): where E is energy, m is mass, and c is the speed of light. If light speed is reduced to nothing, then the square of nothing is nothing. Therefore, energy equals mass. Using thermodynamics (heat transfer), Hawking found significant energy emanating from black holes. In other words, they are in entropy (decaying) by giving off energy.

Mysterious Ideas

We *Homo sapiens* love the mysterious. We demonstrate this every day by being captivated by the fiction and drama of our imaginations. Look at the current sensations of the silver screen: vampires, *Star Wars*, *Star Trek*, Marvel Comics, dragons, and futuristic civilizations. Historically, in addition to being evangelistic, mystery was the foundation of the Jesus Movement in Judaism—the ability to walk on water, make many loaves and fish from only one, healing, raising the dead, making the blind see, and so on. Of course, this became secondary when the Roman conquerors adopted the religion to solidify the power of the state.

So, it is today with science. We have those who would tell you about extraterrestrial beings. While there is evidence that numerous other planets exist with similar conditions as Earth, there is absolutely no evidence that beings from those places would ever come into our presence. Why would they? Why haven't they been

---

[66] Hawking radiation reduces the mass and energy of black holes and is therefore also known as black hole evaporation. Because of this, black holes that do not gain mass through other means are expected to shrink and ultimately vanish. Micro black holes are predicted to be larger emitters of radiation than larger black holes and should shrink and dissipate faster. https://en.wikipedia.org/wiki/Hawking_radiation

observed by masses of us? What would they gain if their technology is so great that they can travel greater distances than we can conceive? Obviously, this is opinion, not science.

When it comes to extraterrestrial life, there is bad science, conspiracy theory, and fantasy. Any extraterrestrial life form advanced enough to travel though light years of space would have nothing to learn by probing the biology of any life form here on Earth. As for time travel—the ability to go forward or backward in time—it has not been demonstrated by the fact that we would be overpopulated by those coming back in time to see us in our current time zone. The fact that we have not had one simple occasion of this is proof enough that time travel has not been discovered in the future.

Put these concepts together with astrology from the Babylonian empire, which still is practiced today, in which it is believed that God speaks to us through the stars, and consider it all to be great fantasy or a good myth. Like the vastness of the universe, our minds can create endless nonsense that implores us to wander into creative and imaginative spectacles.

To permit ignorance is to empower it. To do nothing about it—such as a few philosophers, political leaders, and theologians do, should be a crime. Letting pseudo-scientists proclaim absurd ideas should also be a crime. So, should letting all ideologies, including our schools and religious institutions, teach our children untruths. Not until we purge this superstitious thinking can we embrace all that our minds currently have to offer.

Science without religion is lame, religion without science is blind.

Albert Einstein

Stop, Think, Discuss

Homo sapiens have been given great insight and, like Newton's third law (to every force there is an equal and opposite force), evil can also usurp creative thoughts and use them in an opposite way for power and/or financial gain. Can you think of examples?

What is your opinion of extra-terrestrials? Or UFOs?

Can scientific law also be used in a theological sense, as I have done with the suggestion of Newton's third law?

Question one: In the discussion of dark energy and matter, is there a divine intention of God and what is the understanding of man? Is God even involved? Is this God's creation or man's?

Question two: When a mind like Einstein's is considering the probabilities of space as he did, what do you think he used to control his thinking to see clearly the issues and not the surrounding drama or nonsense? How is it possible for a human to consider deep thoughts outside normal parameters that could easily be considered mental illness in another age?

Question three: We discussed the stars, dark matter and dark energy, mysteries and Brainiacs. What appears to be significant in this? What part of this story do you consider a Godly creation and what part from the hand of man?

Question four: Absolutely everything that happens has a precursor that makes it happen. Then the event happens. Could you have determined, before the event, the cause of the event? Explain the expanding universe, as discovered by Hubble.

Question five: How would you describe this chapter in your own words? Do you see logic or speculation within it?

Question six: Are there issues in these statements with which you disagree?

PART FIVE

THE FRONTIER OF DISCOVERY:
SCIENCE AND THE QUANTA

*How many people over the years have confused theology as
science, and more recently science as theology?*

Chris Pedersen

*To learn, one must not only read but must also think.
Thinking is best done through discussion.*

Chris Pedersen

Stop anywhere it would be prudent to ask any question.

CHAPTER 34

Understand the Difference:
Science and Theology

Science

Typically, modern science is subdivided into:

1. Natural sciences, which study facts, matter, or substance;

2. Social sciences, which study people, politics, and societies;

3. Formal sciences, like logic, mathematics, and theory of computing.

Formal sciences are distinguished from the observed or experienced (empirical) sciences because formal science is not dependent upon observation.

Disciplines that use science, like engineering and medicine, are also called applied sciences.

True science is related to research, proofs of the research, peer review, and publishing results without prejudice toward any point of view. This is significant this day and age. It is a way of qualifying results in the public eye as to the integrity of the results.

I suggest that published scientific results that offer an opinion or point of view, which rarely get published in recognized journals, and that when published go through poor or no peer review, would be more correctly called "political activist" science.[67] Such scientific results are often produced by scholars who are more interested in a personal set of values that is prejudiced toward their point of view. Just as theology historically presented a point of view that represented theology and (incorrectly) interpreted science, so today we are beginning to see science that represents a political point of view to justify an opinion. We humans are quick learners, especially

---

[67] John Bohannon, a biologist and science journalist based at Harvard University wanted to test if the scientific journals that charge a fee to publish articles had adequate peer review to weed out shoddy science. Of the 300 or so journals he approached... only 40 per cent rejected the paper. See the full article at https://www.cbc.ca, "Why a Harvard scientist wrote a bogus paper and submitted it for publication"

in human dynamics. Some of us have learned from history how to manipulate public opinion using the currently popular medium or culturally accepted criteria of the time. Since science has become the popularized "truth", we have this new phenomenon of political activist science. Recently, I was able to listen to the dissertation of a university professor who dissed his own peers for this very distasteful lack of integrity in their science.

From ancient times through the nineteenth century, science was more closely linked to philosophy than now. Today, in the West, the term "natural philosophy" refers to a study of philosophical components of science such as physics, astronomy, and medicine, among many others. Historically, in the seventeenth and eighteenth centuries, scientists were pointedly gathering knowledge in terms of the laws of nature. The word "science" from that time slowly changed to the term we know as the "scientific method", which is a structured, non-opinionated way to study the natural world. We are back to that word "ontology", again.

For a scientist to be believed when they publish their work, they must have integrity. This, in ethics, means, is regarded as having honesty and truthfulness or accuracy in one's actions. So, to *trust* scientific results, *truth* and *honesty* are the glue that binds integrity.

Because some of the reality of some of the new scientific culture, which is that some science is becoming more like religion than science, we are going to explore our structure of science to the last detail. Like the eagle that flew in an Archimedean spiral, instead of a circle, until he disappeared up his own anus.

Let's develop a dialog toward a comparison of science and theology.

What are Truth and Integrity?

Truth[68] is what you believe you know or perceive with your eyes. Just ask any police constable investigating a vehicular accident, interviewing witnesses. Everyone tells a different truth, and all believe they are correct! In ancient Greece, truth was what was decreed to be true by a panel of wise scholars. If the scholars recorded that a horse had seventy-two teeth in its head, that was the truth, regardless if you were a farmer and knew that when you counted the horse's teeth, there were only forty in a male

---

[68] Oxford Dictionaries© Oxford University Press—the quality or state of being true. "He had to accept the truth of her accusation."

horse. In slightly less ancient Rome, truth was what was consistently done by others before you. This is like your mother telling you to put a little salt into the potatoes when you cook them because she learned that from your grandmother who learned it from your great-grandmother who in turn learned it from ... You see the train of thought? Another good example is the Christian Bible (written two thousand years ago in Roman times) where pages and pages of dialog "prove the truth" of who begat whom from before the dawn of the planets. If you went to court during these ages, the accepted truths would be these historical methods.

Real scientific truth is the published result of experiment that has been proven and reviewed by a peer group. Then we accept whatever it is as the truth.

Today, truth begets trust and is considered a bedrock of democracy. Truth is expected, and when it is not apparent, the person not advocating the truth is branded forever as not having integrity.

When truth as we know it in the Western world is lost, trust goes with it.

Trust is a basic building block of not only science, but also in democracy and capitalism. Of course, capitalism and democracy are co-joined twins; failure of one determines the fate of the other. Perhaps this is a possible correct statement of science and theology, also?

A scientist whose results cannot be trusted is not going to impact the scientific world in any meaningful way.

Here is a scientific equation of my own making. It is the equation of a significant character assessment. where honesty is a major component in integrity. ((Honesty + Truth) = Trust) = Integrity

Integrity[69] in Latin would mean whole or complete, and our Western culture would say it is a personal choice of having good morals; ethics; consistency of application of these values including honesty, truth, and trust; obedience of laws, both legal and societal; ability to accept fair criticism, admit errors, and attempt to correct results when mistakes are made. As you can see, integrity is a subjective decision, not a scientific one, which in Western society indicates the values espoused by theological statements, particularly those of Judaism and Christianity. Integrity is also a good proof of theological input as a valuable building block within human culture. The

---

[69] Oxford Dictionaries© Oxford University Press—the quality of being honest and having strong moral principles; moral uprightness. "He is known to be a man of integrity."

best societal example of integrity for me is a trip I had in Japan. A traveling companion departing the subway left a very expensive camera on the seat. They immediately went to an employee at the station to report what had happened and request assistance. The employee looked at his watch and said that the train would be back in about 32 minutes, and the camera would be on the seat where it had been left. That is exactly what happened. The natural assumption by everyone in Japan was that would obviously be the case. Cultural integrity.

I have seen in other cultures where integrity applies solely within the family unit, and not for or to greater society. In these places, the camera would not be there. The person taking it would see it as a gift from some unseen deity for the benefit of their family. But this same person within the family unit would have family integrity and never harm, discredit, or steal from family.

A great scientist sees integrity as their treasure. They may not be rich or well-known, but they have integrity. They work hard at being perceived as having integrity. They will and must guard their integrity with great care. Without integrity, they would never be considered believable, any results would be suspect, and their peer group would not even review their work. They would never be published in the important and recognized scientific journals.

Philosophy

Philosophy is a discipline with a core structure of logic, aesthetics, ethics, metaphysics, and epistemology. Metaphysics are a division of philosophy, concerned with the fundamental nature of reality, which requires ontology and includes cosmology. It is the theory of knowledge, its methods, validity, and scope. Epistemology is exploring the distinction between belief and opinion, while ontology is about what exists without injecting personal bias or prejudice.

Philosophy is the study of ideas and concepts without the use of data. Philosophy, science, and theology are conjoined triplets, a three-ring circus of never-ending discussion. Each part has opinion overlapping the others in effort to gain sway or authority over the whole. While each part identifies as singularly independent; they are in fact like all parts of the universe, necessarily dependent, each upon the other, through natural cohesion. Read this again, it is significant.

Theology Revisited

Theology is the study of the nature of God and religious belief. Christian theology is simply an attempt to understand God as God is revealed. No theology will ever fully explain God and God's ways because God is infinitely and eternally higher than we are. Theism is a belief in the existence of a god or gods, especially belief in one God as creator of the universe, intervening in it, and only sustaining a personal relation to his creatures through original creation, not in everyday relationship.

From the dawn of *Homo sapiens,* the principles and practice of theism has been a building block of civilization. It has made possible the ability of people to live together in groups of more than the family unit. It has provided mystery to the mindful needs of our forefathers, who were searching for truth. It helped create civilization with an outcome that usually included love, forgiveness, charity, and grace. Unfortunately, this was not always the case. Because theology includes historical writings, and some of those historical dramas were attempts to reconcile the mystery of the world, theology was often historically accepted as science. Theology is also future thinking and eclipses every cause-and-effect thinker. The ingredient that provides us this future thinking is faith.

As we have seen in prior chapters, the push and pull of human endeavor also includes powerful emotions and vices of greed, lust, avarice, pride, and envy. There is also what I would call "God's third law"—to every good thought there is an equal and opposite thought. Thus, another absolute; but only in my opinion, based upon the science of Newton, but not science.

And so, we arrive at our verbal conclusion for the conjunction of the subjects of philosophy, science, and theology.

Theology is not science.

And

Science is not theology.

If theology and science are oil and water, then philosophy is the jar that contains them. You can shake the jar to make an emulsion, but the oil and water will always separate. However, together, all three have shaped our world and are vital in this act of shaping it.

Stop, Think, Discuss

Try to make a list of all the things that come to your mind that confuse science and theology.

Did you discover that this list is mostly theology trying to be science? Could that be because there is a lengthy history of theology as science?

Can you make a list of science that would have much different or positive results if theology was the moderator of it? Like eugenics?

Think about science that requires faith to believe. Is it really science or is it in fact theology? This will also be discussed later.

Question one: How does God enter this discussion of scientific reasoning? Is God even involved? How did this discussion even enter the mind of humans, and why only humans on this planet?

Question two: What is the *ontology* of this discussion? That is, what is the actual information, free from our personal bias, of these statements? And what is the *epistemology*? That is, what is the knowledge, as learned by us, including our preconceived notions, of these statements?

Question three: What is the more powerful discussion in this chapter? Is it theology created by God? Theology created by man? Or science discovered by man?

Question four: Absolutely everything that happens has a precursor that makes it happen. Then the event happens. Could you have determined before the event, the cause of the event? Also, could the event have been prevented if the cause had been known beforehand? Do these statements have any relationship to this chapter?

Question five: How would you describe the idea of interrelated science, philosophy, and theology in your own words?

Question six: How do you feel about the creative thinking within the process of thought in this chapter? Provide your opinion to others with explanation.

CHAPTER 35

Science That Isn't

We have discussed the conflicts of science and theology (or theism) at length in previous chapters. We have demonstrated how we basic humans, who are just going along for the ride, maintain a posture of neutrality for a long time, but then rebel against those who distort truth, trust, and honesty for other purposes.

Roman culture through the actions of Constantine absorbed the Christian Church, giving it power and riches. Eventually this created have and have-not people in society. The Renaissance and Protestant Revolution attempted to correct this in the 1500s. Darwin and Mendel gave us the precursors to DNA medicine in the mid-1800s. Galton, a cousin of Darwin, usurped this and gave us eugenics. By the mid-1900s, eugenics resulted in sterilizations, electroshock therapy, taking indigenous people from their homes into residential schools, and finally, with Hitler, genocide.

Potentates, kings, tribal leaders, dictators, theocracy, bishops, popes, and Indian chiefs have all been implicit in controlling our lives, often with the power of theological interpretation to justify their power over us.

Times have changed. We have lost empathy for the power of these historical platitudes. Today, we search for personal satisfaction, for justified authority to rule our society. Many have left theological history and have become "spiritual". Spiritual seems to have become an antithesis to formal religion and a way of expressing neutrality to atheism, while at the same time believing it expresses some form of acceptable analogy to Christianity. In postmodern society, that is anyone born after Elvis Presley; if it feels good, it must be good. In the modern era, the hundred years starting in about 1850, we use science for our justification, and in so doing, we create another form of religion, this time science-based or science-justified. This is the reason behind the long diatribe at the beginning of this segment of the book on truth, trust, honesty, science, and the distortion of power.

Bad Science

News flash, not all science is good (or even science). We want building blocks to continue to improve society and are asking science to deliver those building blocks. As always and as history proves over and over, there are those who will subvert good ideas into a negative force in search of prestige, power, and riches within this modern scientific movement. Those scientists create a form of junk science, which is what I call science that is not value-based, or which has not passed peer review. For-profit publications offer this science a platform to publish reports that often stretch the credibility of other scientists who work hard to stay within the bounds of pure science. Often, to support this departure from reality, these scientists also use junk statistics in their manuscripts to justify their political activism. Sadly, this is not restricted to one branch of science; it has become universal in application to all branches of scientific discussion.

We must learn how to discern good science from junk science. We need to be able to determine ourselves what is bad science and what is good.[70] Also, if it appears to be bad science due to political or other influence, is it really bad science? Ditto for apparent good science. The field of choices for me to present this discussion are several. I have considered the use of vaccinations to prevent disease and the backlash of those who oppose vaccinations currently. Then there is Genetically Modified Food, and the large following of people who think it is a distortion of the truth to say GMO (DNA-engineered food) is safe while they believe it is not. I have chosen climate change because of the controversy within this field and the very large component of people who are directly involved. Meteorology is a field I know quite a bit about, and climate change is under this category.

So, let's look at the current contention of global warming. I would call this computer-based modeling. Environmental study (science) used to be the study of meteorology, and a field within which I spent a whole lifetime studying, teaching, and working. Meteorology, historically, was the forecasting of weather using data

---

[70] https://www.psychologytoday.com, has an excellent article on "How to Tell the Difference Between Good and Bad Science"

collected by manned stations where they measured temperature, wind, pressure, clouds, visibility, and humidity. It was and still is primarily used in aviation but is obviously also used in other occupations as well. Trucking, farming, hydro, and even the public have a need to know what the weather is going to be to the best accuracy that can be forecast by meteorologists. In the 1970s, meteorology converted to gathering data from satellite images and machines, as well as manned stations, which have gradually been eliminated. The images collected, especially from satellite, were then computer modeled as to what historically happened to weather when that pattern was observed. From that computer data, a forecast was made, which originally was abysmal compared to the forecast produced from data of manned stations.

However, the accuracy has improved over time, but it is still only a best guess based upon historical computer data of what happened prior to this satellite image of weather. I must say that upper weather data—wind velocity, direction, and turbulence at altitude—has always been very accurate, with or without computer modeling. Computer modeling requires the presumption that whatever happened before will happen again, if the data and circumstances are the same. Accuracy in surface weather forecasting has never been accurate beyond three to five days, and even then, its accuracy is dependent upon the area where you are forecasting weather. Oceans, mountains, and valleys influence weather in ways that often prevent accuracy beyond the immediate.

With computer modeling, the only way to predict future events is to accurately record all historical events as they happened using the same criteria to portend the future. It is imperative that all data affecting the result is included in the calculations, or your prediction will fail. Anything affecting the weather pattern outside the normal historical value will ultimately create an incorrect forecast, but then that data will be in the computer for the next event.

Computer modeling is not science; it is data collecting and using that data to influence future thinking. Computer modeling does not account for volcanic eruptions, earthquakes, or tsunamis. Nor does it account for particulate suspended in the air that has been put there by exceptional conditions or catastrophe. All of these affect the outcome of a forecast and cannot be predicted.

For some people, global warming prediction, recently changed to climate change, is already a religion. For others, it is something to be aware of, and of course for even others, it is a sham or possibly junk science. What exactly is going on with the science of global warming? What are we being led to believe and who are the authors? Is the science intrinsically correct or partly correct? When taking a statistical review of heating the Earth, are the datums for starting point and ending correct, or prejudiced to give false viewpoints? Do the proponents have integrity? Does their science have integrity? Who are the antagonists, and do they have any integrity? There are so many published environmental scientists favoring this. Can they be correct just by their numbers alone? Does either side of the argument have anything to gain? Since dire future predictions are based upon computer modeling, has this computer model been correctly defined? For instance, did computer gurus correctly calculate the effect of the radiation of the sun being absorbed by the earth, and who proofed their results? These are questions each of us must ask to ascertain the value judgment of the science of climate change.

There is something quite wonderful happening from this new scientific and religious discussion. We are becoming better custodians of our planet. We care. In theology, we are called to do this, and it has been a struggle in the past. Now we recycle. We care for the other living things within our environment, from trees to beasts to bugs. Science has been pushed to invent the LED light, electric vehicles, and solar power. We are trying to clean up our past environmental misdemeanors. And it is good.

But we are insignificant in the construct of the universe, even on this planet. Everything we do is reversed in an instant with the eruption of one large volcano or the arrival of one large piece of rock from outer space. Our footprint is surely all over the world, and it is affecting the conditions upon this orbiting vessel, but we are not in control. We are not gods and not in control—yet! Even if we think we are.

Let's consider the exercise of us less-educated folks forming an opinion on global warming. This is just an exercise of thinking not intended to change opinions already fixed.

Carbon dioxide is increasing in our air on the planet, and the result is increasing temperatures. Legions of protesters (often acting like extremists and disconnected from economic reality) are out in force to demand that we discontinue the use of carbon-based fuels, which produce CO_2 in their burning. They are even purporting that oil from one region of the planet is dirtier than from another region. While on the one hand this is a great custodial cleanup effort, and we should be trying to invent a cleaner and more manageable fuel, on the other hand, increased CO_2 levels also increase the growth of green vegetation, a lot of which is food. It is used extensively in green houses to grow our vegetables. Green vegetation cleans the air of impurities including CO_2. Lastly, a large producer of CO_2 is the human body. Nearly forty percent of our human-exhaled gas is human-generated CO_2, a byproduct of O_2 mixing with sugar in our bodies to make energy. Humans produce more CO_2 than all the passenger vehicles on this planet, as stated in an earlier chapter. So, what is the truth? Does the scientific hypothesis include all data, or does it pick and choose only selected data to make a political activist statement? And what should we do about this in our selective analysis? Genocide is not a reasonable thought process, so what other options are there?

Narcissism and Science

A local doctor (implication scientist?) wrote a paper that stated refueling an ocean liner at sea, using floating buoys as protection to withhold oil should there be an accident, was an environmental disaster waiting to happen. Investigating the credentials of the scientist turned up the fact he was a dentist masquerading as a scientist with an interest in writing just to be acknowledged in the media. We peons need to become discerning through our questions about results that require faith to understand.

The Exercise of Rising Sea Levels

There is a huge number of people who believe that there is going to be at least a nine-foot rise in ocean levels due to global warming. Intellectuals state that melting ice on the mountains, Greenland, and portions of the South Pole will be the cause. Other pseudo "intellectuals" state that melting sea ice will also add to disaster, and therefore, we will see a tremendous rise in sea level.

Let's try to be objective. It is not stated that the ocean floor is rising to displace huge volumes of water onto the land mass. The only statement is that melting ice will cause a rise of nine feet (3 meters) of the water that covers more than seventy percent of the planet. Remember, sea ice has already displaced the amount of water it will occupy, so melting floating sea ice is ineffectual.

Let's practice the discernment required to get to the truth and trust the conclusions wrought by our activist environmentalists. Here is the exercise for you to reason out the rest of this statement for yourself. Use logic and reason as you have intuitively inherited it.

The ocean occupies more than seventy percent of the surface of this planet. Also, ice has a volume seventeen percent larger than water due to decreased density as it freezes, which is why ice floats. For seventy percent of the planet (the ocean) to rise nine feet, there must be enough ice on the other thirty percent (land) to produce that water when it melts—seventeen percent more ice than if it was water in volume.

So, do the math: Thirty percent of surface (land is 1/3 of the surface area of the Earth) must have enough ice to melt into nine feet of 2/3 of its surface area. Thus, nearly twenty-five feet of ice must cover all land f it were uniform throughout. (including seventeen percent for change of state). This is nearly triple the height of the necessary resultant water required, including seventeen percent to melt, to increase the ocean by nine feet. That is, every square inch of land space will have to be covered by more than twenty-five feet of ice. That is a lot of ice, if it were to be distributed evenly across the land surface. But it is not distributed that way, so we need to hold on to our initial expectations of "not possible" until we think further, like a scientist.

Where is the ice and how much is there? Well, Antarctic and Greenland contain the most ice. I can say this because I have seen it with my own eyes. My eyes also told me that the surface area of Greenland was not large enough to contain that much ice. But let's do the math! A whole lot of the Antarctic ice is already in the water, but not all of it, and the continent is very large, so it would be a guess on my part as to how much land-based ice is there to melt. With reasoning, it is also very environmentally sheltered and not likely to all melt in several thousand years. Greenland is a known quantity. The Danish government has measured the ice cap on Greenland very accurately to a volume of 2,900,000 km3. The surface area of Earth is 510,100,000 km2, and only seventy percent is water, or 357,000,000 km2. Do the math; make cubic km into area one meter deep and divide the resultant into the wetted surface area of the Earth to get a result that should approximate the height of water from the melting of glaciers based only upon Greenland. I get 6.74 meters (including the reduction of seventeen percent), which is a lot of water. Since seismic survey is an old science and can be trusted, we must assume that sea levels will rise if Greenland melts. Don't forget, if it is warm enough to melt Greenland, then Antarctica will be melting too! And I get to eat crow because I would not have believed it using my eyes only as the resolution to this calculation.

Scientific conflict arises. Others try to contrast this argument with the idea that as the ocean increases in temperature, it will expand, thus causing an additional rise in sea level. Real science states that ocean water has a "thermocline" in it averaging approximately one hundred (or less) meters down from the surface, beyond which sunlight, waves, and current do not affect the temperature of the water.

Thermocline in the Ocean

Most of the shortwave radiation energy of sunlight is absorbed in the first few centimeters of the ocean's surface, which heats during the day and cools at night because heat energy is lost to space at night by long wave radiation. Wave action mixes the water near the surface layer and distributes heat to deeper water, resulting in a temperature that may be relatively uniform in the upper one hundred meters. This depends upon wave action and the existence of wind and currents. Below this layer, temperature remains relatively stable over day/night cycles. The temperature of the deep ocean drops in an adiabatic way with depth. The temperature well below the surface is usually not far from zero degrees C.

Therefore, mathematically, find the rise in level of water if the ocean is heated to a depth of one hundred meters uniformly by fifteen degrees C (a lot) and the coefficient of expansion of water (0.000214 per degree C) is used to find the increase in volume. Doing the math, the water will increase in volume by a total of 23.5 cm (just under ten inches for three hundred feet—one hundred meters—of water).

How do you reconcile these calculations from the political activist science? What is different between the calculations? Are both calculations provided and, if so, are they done in a way that is easily understood? The calculation proposed here is available to anyone using the Internet (it is not quantum physics). Has one calculation used criteria different from the other and if so why? Could one just be a political process to justify something, or a mistake? Have the datums been established correctly in either calculation? Do the persons making either of the claims have integrity? Does the resultant try to make a point that would change something in a way unrelated to the calculation? Science provides data, and politics provides opinion.

Sarcasm As a Point of View

Then there are the historical references that would be used to comment upon current scientific thinking. These are used in a sarcastic way to debase or cloud current scientific thinking in a way that would lead one to believe that rising sea levels by global warming is a sham. By example, this is the most recent email I received from a relative with a newspaper article;

> The Arctic Ocean is warming up, icebergs are growing scarcer, and in some places, the seals are finding the water too hot, according to a report to the Commerce Department yesterday from the Consulate at Bergen, Norway. Reports from fishermen, seal hunters, and explorers all point to a radical change in climate conditions and hitherto unheard-of temperatures in the Arctic zone. Exploration expeditions report that scarcely any ice has been met as far north as 81 degrees 29 minutes. Soundings to a depth of 3,100 meters showed the Gulf Stream still very warm. Great masses of ice have been replaced by moraines of earth and stones, the report continued, while at many points well-known glaciers have entirely disappeared. Very few seals and no white fish are found in the eastern Arctic, while vast shoals of herring and smelts, which have never before ventured so far north, are being encountered in the old seal fishing grounds. Within a few years it is predicted that due to the ice melt, the sea will rise and make most coastal cities uninhabitable.

> I must apologize. I neglected to mention that **this report was from November 2, 1922 as reported by the AP and published in** *The Washington Post* **96 years ago.** This must have been caused by Model T Ford emissions or possibly from horse gas.

(Sic)

Another viewpoint of junk science is that if the calculation is proven to be outside reality, then another reason might be provided. Something like the suggestion made earlier that it would be due to rising of tectonic plates on the sea floor. If this was the case, does it pass the smell test? As Kepler would say, "Something is rotten in the state of Denmark!" (*Hamlet* I.iv.90).

As an aside, there was one pseudo intellectual (activist environmentalist) in my area who tried to convince us that water would be higher in some places than others. I asked if he was talking about tides, and the response was "no". He believed that water would suddenly undo the laws of gravity and seek its own variable height. Well, except for tides caused by the moon, we should all know that water seeks a uniform level. We even use this leveling capacity with instruments in construction to level the foundation of a new building. When science requires faith to become believable, then it becomes religion.

What is Science?

So, what is science? Is it the new religion? Does science have all the correct answers? Will someone with malevolent thoughts of power and greed take the genius of DNA discoveries and turn them into a revolting world of disaster like Hitler did with eugenics? Will the experience of politically correct thinking take over religion and modify civilization with a pent-up vigor toward science being omnipotent?

I suggest again, science is not religion, and religion is not science. They are oil and water. Both are compatible and should be symbiotic. One should support the other and neither should conflict with the other. Theology influenced our thinking of science for 1,500 years; perhaps a scientific modification in theological thinking for modern scientific times is possible?

Stop, Think, Discuss

Many countries are placing a carbon tax on their citizens, stating it is to encourage reduction of burning carbon in the atmosphere. This tax resembles the indulgence tax from the Roman Catholic Church between 1400 and 1500, a payment of forgiveness for your sins, effectively reducing your time in purgatory.

Do you see how a tax on carbon is going to do anything other than assuage your guilt?

Question one: In these statements, what is the divine intention of God and what is the intention of man? Is God even involved? Is this God's creation or man's?

Question two: What is the *ontology* of these statements? That is, what is the actual information, free from our personal bias, of these statements? And what is the *epistemology*? That is, what is the knowledge, as learned by us, including our preconceived notions, of these statements? Use the calculations for reasoning.

Question three: What appears to come first in this? Was it an idea created by God or by man?

Question four: Absolutely everything that happens has a precursor that makes it happen. Then the event happens. What is the reason behind global warming as the facts of measured increases in temperature are provided?

Question five: Describe the interplay of science and theology and the interference of human endeavor.

Question six: Are there issues in these statements with which you disagree?

CHAPTER 36

Back to the Future

The science of the atom: Studying the small, very small, and minuscule particles of matter. In other words, looking at the quanta or composition of the universe by atoms. Remember, except in literature where a quantum leap is a large leap in judgment written as a thought upon a piece of paper, the quanta or quantum in science is the study of infinitely small particles of matter that compose all the universe.

There are also some futuristic ideas currently being explored by scientists about the future of the universe—or possible future of the universe. At this juncture, everything is conjecture, an idea, a clever thought process being explored as to its reality. All scientific study begins with ideas and thoughts that may or may not become reality when exposed to the rigors of physics and proofs with peer review.

This begins in Greece in 450 BCE. Leucippus, the teacher, and Democritus, the student, created an eclipse of thought that converted scientific thinking for all time. Democritus stated that everything in this universe is made up of endless space in which innumerable atoms run. Space is without limits, has neither an above or below, and is without boundary. Atoms have no qualities other than their shape. They have no color, weight, or taste.

Atoms are elementary grains of reality and cannot be further subdivided, and everything is made of them. Moving freely, they collide, catch and push each other. Similar atoms attract each other and join. Your thoughts are thin atoms, your dreams are atoms, the light you see is atoms. They make up structures, and we, like the animals, are made up of atoms in this continuous march. This happens over the course of eons of time. So said Democritus in dozens of books.

All the books have been destroyed. Only the words of others quoting and summarizing him are still available. Aristotle was aware of Democritus and spoke of him with respect but fought against his ideas. Plato, student of Socrates, was also aware but was not so polite in his rebuke. We know from astrophysics that the Aristotelian and Ptolemaic theories of the universe held sway with the Christian movement of the Roman Christian Empire. This is undoubtedly why the books of Democritus were destroyed—they didn't conform to the values of the Christian Roman Empire at the time.

The idea that the atom was the last finite or smallest particle of matter that you could divide anything down to was also made. It was stated that you could take a raindrop, divide it into two raindrops, and then four, and so on, until you got to a point where you only had a molecule, and you were done. It wasn't until the early 1900s that this idea became reality for the scientific world. In 1905, Albert Einstein defined atomic hypotheses in mathematical terms and determined the size of the atom—2,205 years after Democritus was able to make this connection.

What reasoning, what process of thought, could make one individual in a world of people just recently out of Africa in their migration as hunters and gatherers, so impossibly advanced for the age?

Part of that answer is the use of mathematics.

To summarize the discussion on this, mathematics as we know it was given to us by Pythagoras (570 BCE to 495 BCE). Pythagoras gave us numbers and uses for the numbers through calculations in his school in Italy after traveling extensively from his home in Greece. Aristotle gave the study of mathematics a new perspective through his book *Physics*. That is the very word origin that is used today in the study of physical questions and mathematical resolution of problems.

Do not say a little in many words but a great deal in a few.

Pythagoras[71]

Plato refined Pythagoras' use of mathematics into the language that is used to understand and describe the world. Plato taught Aristotle. History says that Plato carved onto his door at the school, "Let no one enter here who is ignorant of geometry." Plato is the reason for the success of Western science in the modern age.

[71] Pythagoras of Samos was an ancient Ionian Greek philosopher and the eponymous founder of Pythagoreanism. His political and religious teachings were well known in Magna Graecia and influenced the philosophies of Plato, Aristotle, and through them, Western philosophy. en.wikipedia.org

But it took more than a thousand years to reaffirm the need for science and discovery in the Western world. Ptolemy, an astronomer in the first hundred years of the modern age, was the last of that age to write correctly and precisely about the random movements of the planets. It was Earth-centered and acceptable to the new Christian Roman power, and therefore, acceptable in that genre. To the eye, it was correct, but the true test is the mathematical solution. Obviously, it was very incorrect, but it took more than a thousand years to discover this.

The science of observation of the stars became the venue of Indian, Arab, and Persian observers. This, too, was usurped by religion when in 1100 AD the Islamist school in Bagdad was swayed by the opinion of the scholar Hamid al-Ghazali, who wrote a series of papers questioning the philosophy of Plato and Aristotle and declaring mathematics to be the philosophy of the devil. He declared that it was more important to study the Islamic religion than science.

Finally, in the Western world, in 1543, Copernicus dared to append the Pythagorean/Ptolemaic ideology, thoroughly improving it. His lust for the truth was held in check by the threat to his life should the authorities rebuke him. So, it was on his deathbed that he published his corrected version of the Ptolemaic evaluation. The Danish school of Tycho Brache supplemented the Copernican model and produced a brilliant scientist named Kepler, who calculated the movement of the planets as elliptical not circular.

Then, Galileo made observations using a telescope for the first time to certify these findings. But the influence of religion was in jeopardy within society, and the religious science was being challenged. So, there was one last effort to re-certify the authority of Christian scientific belief, and Galileo was the collateral damage. However, this was the last vestige of the Church resisting science, and history would now proceed with mathematics, physics, and science in pretty much open-ended progressive use, which was even acceptable to the Church, as we understand in modern times.

Galileo is also the first to do an experiment of mathematics other than calculating the position of the stars. He was the first to study falling objects. He discovered that objects accelerate as they drop; something heretofore unknown. He determined that they accelerate by 9.8 meters per second per second. Suddenly, science is also the purview of the Earth and the stars.

Science is now relevant in 1610 AD.

Stop, Think, Discuss

Think of mathematics as a Rubik's Cube—complicated, three-dimensional manipulations that, if done correctly will result in an order obvious to the eye of the beholder. Some of these scientists can see numbers like this without even putting pen to paper.

Do you know anyone who can use their mind to turn what would be chaos into organization, to the great amazement of all?

Question one: Where is God in all of this? Is God even involved? Is this God's creation or coincidence from man forming random thoughts into creative ideas?

Question two: What is the *ontology* of these statements? That is, what is the actual information, free from our personal bias, of these statements? And, what is the knowledge, as learned by us, including our preconceived notions, of these statements?

Question three: What appears to come first in this? Was it an idea created by God or by man?

Question four: Absolutely everything that happens has a precursor that makes it happen. Then the event happens. So, what is it that came first in this discussion? Thinking about all the events in this discussion, pick one. Could you have determined, before the event, the cause of the event? Also, could the event have been changed if the cause had been known beforehand?

Question five: Provide you thoughts on this chapter. Explain.

Question six: Do you agree with the statements in this chapter or is it fiction in your opinion?

CHAPTER 37
Newton Arrives and Changes the World

Then, in the mid-1650s, Sir Isaac Newton arrived with a mind for science unlike any other modern human experience to date. He is often called the most brilliant scientist of the modern era.

Newton published a book called *Principia*, which founded modern science. He asks you to imagine that Earth has many moons, and in particular, a "little moon". It orbits the Earth just above the mountaintops. He deduces the speed that it must have to orbit the Earth. Kepler's law indicates the speed of the orbiting moon is relative to the distance of the moon from the Earth, including the time to complete the orbit. Using a comparative analogy to our real moon, he determines that the little moon would take an hour and a half to complete its orbit.

Since this moon does not go in a straight line, but orbits in a circular trajectory, it continually is changing direction. A change of direction is called "acceleration". The little moon must accelerate toward the center of the Earth. Newton did the calculation and discovered that this number "the force of acceleration of orbiting bodies around the Earth" is 9.8 meters per second per second, the same number Galileo discovered for falling bodies!

Simple cause-and-effect thinking again: If the effect is 9.8 then the cause is the same. Therefore, the force that causes the little moon to "accelerate" around its orbit must be the same as the force that causes objects to fall to the Earth. Gravity! Without this gravity, the moon would leave Earth in a straight line. So then, also, the real moon must be experiencing the very same forces of motion, and the planets that orbit the sun as well, and so on.

Suddenly the hypothesis is that the universe is a large space whereby the bodies attract each other by means of forces, and there is a universal force, gravity. Each body mass attracts every other body mass.

Suddenly, a great visual perception was experienced. There was no longer separation between the heavens and the Earth. Aristotle's natural level of things was no longer correct. Anything set free from the attraction of gravity is set on a straight course and will continue that line forever.

Newton set his mind at furthering his exploration of the subject by trying to determine the force of attraction over distance. He came up with "Newton's Constant" using the letter G for gravity in his calculations. It causes things to fall on Earth and to orbit bodies of matter in the universe, and it is the same force.

The universe is suddenly an infinite space of stars without limits and without a center. Inside this space, massive bodies run free and straight unless they are affected by the force from another body, which deviates them. The wisdom of Democritus had resurfaced, and indeed Newton studied the concepts put forth by Democritus two thousand years after they were first postulated. The world was now re-experiencing the world of Democritus, but with mathematics to prove it.

The Newtonian intellectual framework, meaning observed and mathematically proven, is beyond all expectation. The entire technology of the nineteenth century and to this day rests upon the formulas of Newton. My engineering professors taught strength of materials using his formulas; flying an aircraft, making weather forecasts, predicting the existence of another planet prior to seeing it in a telescope, and the calculations to send spaceships to the moon, are all Newtonian formulas.

Newton conceptualized a new way of thinking: The world is a machine that operates on simple principles, which so raised the levels of enthusiasm that he is a legacy of the modern world. Now, precise equations were available to describe the forces of nature. Calculations that are immensely accurate are now viewed as the fundamental system of laws of the universe. Everything is now clear and known.

Or is it? Looking back, we are aware of a previous statement: "Beware of anyone in the white coats who say they have all the correct answers. For the answers are always changing." We are not always talking about religion. This is also a revelation of science, too.

Newton knew that he could not describe all the forces in nature. There are other forces than gravity that affect bodies. There does not have to be a falling activity to move things; they can be observed to move without gravity.

Surprise number one. Almost all things we see moving are governed by a single force other than gravity: Electromagnetism, which is the force that holds together matter that forms the solid bodies, atoms in molecules, and electrons in atoms. It is the force that makes everything work.

Surprise number two. Understanding electromagnetism requires an adjustment to the Laws of Newton. This adjustment is the birth of the new and modern physics of this day and age.

Stop, Think, Discuss

Newton was a strong proponent of Christianity, but he was not a believer in the Trinity. He felt that the Council of Nicaea in 325 affirmed the Trinity. Newton traced the doctrine of the Trinity back to Athanasius (298 to 373 AD), who said God the Father had priority over Christ. He was convinced that prior to that the Church had no trinitarian doctrine.

From the short review of Christian religions in the first part of the book, which Christian organization also has this anti-trinitarian resolve?

Question one: It is interesting that Newton had a faith when many modern scientists are losing theirs. Do you think Newton's faith is justified? Is God even involved? Is faith necessary in this science?

Question two: What is the *ontology* of these statements? That is, what is the actual information, free from our personal bias, of these statements? And what is the *epistemology*? That is, what is the knowledge as learned by us, including our preconceived notions, of these statements?

Question three: What appears to come first in this? Was it an idea placed in the head of Newton by God or by man?

Question four: Absolutely everything that happens has a precursor that makes it happen. Then the event happens. What was the precursor to Newton's discoveries and was it directed by anything other than coincidence?

Question five: Describe this chapter in your own words.

Question six: Are there issues in these statements with which you disagree?

CHAPTER 38

The Shocking Faraday and Maxwell Insight

This is the story of a Londoner and a Scotsman—Michael Faraday and Clerk Maxwell. Living in London, Faraday became a brilliant experimenter in the nineteenth century. He was able to write one of the best books on physics without knowing mathematics or using equations. He saw in his mind the visions of his experimental world. Clerk Maxwell from Scotland was an aristocrat and one of the greatest mathematicians of the century. The vast gulf of difference of method between them was overcome, and they succeeded in understanding each other. Together, they created modern physics.

Faraday

Electricity of the eighteenth century was not much more than a circus trick. Experiments in the following years to the nineteenth century explored various ways of understanding and using electricity and magnetism. Faraday, knowing about Newton, tried to understand the force of attraction between charged particles and magnetic things. Slowly, and with hands that were constantly in contact with his objects, he saw something new.

He saw a field of forces that is diffused in space and modified by electric and magnetic masses that push or pull the bodies. Much like the way we see iron particles showing lines of force around a magnet, Faraday had the same vision in his head for this imaginary idea.

In the same way that Newton stated there is an equal and opposite force in his third law, and I state the proposal that every positive thought has an equal and opposite thought or action, Faraday saw the magnetic field as having a positive and negative end, just like the north and south points on a magnetic compass. Opposite forces attract, and similar forces repel. Further, distant objects are pulled or pushed directly, but by the action of the forces around them.

This is radically different from the Newtonian idea of a force acting between different bodies. Having introduced this new idea into his thinking and written text, Faraday had discovered that the world was no longer made up of particles that move while time passes. He was conscious of the fact that he was proposing a modification in the structure of the world.

> *No matter what you look at, if you look at it closely enough, you are involved in the entire universe.*

Michael Faraday[72]

Maxwell

Maxwell was transfixed by this thinking and translated these ideas into a page of mathematical equations called "Maxwell's equations", which describe the behavior of magnetic and electric fields. These equations are used to this day in all forms of electronic devices such as antennas, radios, and computers. Not only that, they explain how particles of matter are held together and almost everything we see happening takes place, except for gravity. Everything is described by Maxwell's equations.

Even better, Maxwell's equations tell us what light is. He calculated the speed at which the waves of magnetic attraction move, and the result turned out to be the speed of light. He figured out that light is nothing more than the speed of the trembling of Faraday's lines. So, what is color? It is the speed of oscillation of electromagnetic waves. The faster the speed, the bluer the light; slower is red. Our eyes are tuned to perceive the frequencies of oscillation of electromagnetic waves at different frequencies.

He further determined that it should be possible for the frequencies to be slower than the speed of light. This resulted in the construct of radio waves from the electrical methods devised by Heinrich Rudolf Hertz in Germany. Hertz established the nature of their vibration and that their susceptibility to reflection and refraction were the same as those of light and heat waves. As a result, he established beyond any doubt

---

[72] Michael Faraday FRS was an English scientist who contributed to the study of electromagnetism and electrochemistry. His main discoveries include the principles underlying electromagnetic induction, diamagnetism, and electrolysis. en.wikipedia.org

that light and heat are electromagnetic radiation. With this, Guglielmo Marconi was able to build the first radio.

All modern communication is an application of Maxwell's predictions.

Almighty God, Who hast created man in Thine own image, and made him a living soul that he might seek after Thee, and have dominion over Thy creatures, teach us to study the works of Thy hands, that we may subdue the Earth to our use, and strengthen the reason for Thy service.

James Clerk Maxwell[73]

So, the mind's eye of Michael Faraday, along with the mathematics of Maxwell, is the foundation of our entire current technology. It establishes the fact that it is not necessary to discover empirically the evidence that will further scientific thinking, but that intuition is a larger factor in the evidence of future thought.

The world has changed. It is no longer made up of particles of matter in space, but rather, particles and fields in space.

As the prophet on the street corner would say, "We have all the answers, just turn to XXX and your problems are solved!" Then along came another rather uninteresting and underachiever in school, a boy whose father's name was Einstein.

Ah, but that is another story for another chapter!

---

[73] James Clerk Maxwell FRS FRSE was a Scottish scientist in the field of mathematical physics. His most notable achievement was to formulate the classical theory of electromagnetic radiation, bringing together for the first time electricity, magnetism, and light as different manifestations of the same phenomenon. Maxwell's equations for electromagnetism have been called the "second great unification in physics" after the first one realized by Isaac Newton. en.wikipedia.org

Stop, Think, Discuss

Modern science has now seen a brilliant scientist, Faraday, who was not capable of advanced math or physics, but who still significantly contributed to the enhancement of our knowledge of the creation of the universe.

Do you think you know someone who could give insight to advanced thinking, regardless of their mathematical background? An inventor, teacher, physician, or just a great thinker?

Think of the other end of the spectrum, like someone who is a great thinker, who is also destructive. For example, think of Theodore John Kaczynski, also known as *"the Unabomber"*, apprehended in 1996 after killing three people. Being smart is not always a prerequisite to great thinking.

Question one: Faraday felt and saw in his mind the structure of his electrical genius. So, what is the divine intention of God and what is the hand of man? Is God even involved? Is this God's creation or man's?

Question two: What is the *ontology* of these statements? That is, what is the actual information, free from our personal bias, of these statements? And what is the *epistemology*? That is, what is the knowledge, as learned by us, including our preconceived notions, of these statements? Why would a brilliant scholar like Maxwell consider that a magician like Faraday had any concept of new thinking? What force of nature would bring this odd couple together?

Question three: What appears to come first in this? Was it an idea created by God or by man? What role does coincidence play?

Question four: Absolutely everything that happens has a precursor that makes it happen. Then the event happens. Could you have determined, before the attraction of Maxwell and Faraday, the cause of their friendship? Also, would their scientific discoveries have been prevented if this mental attraction between the two had been prevented beforehand?

Question five: Describe the value of their discoveries in your own words.

Question six: Are there issues in these statements with which you disagree? Has Harry Potter— creative thinking—taken over the dialog on this issue?

CHAPTER 39

Time, Spacetime, and Einstein

Here is an interesting, real-time situation. I am the captain of a modern-day airliner. I am looking out the window for traffic that may affect my flight path. Because my eyes, like yours, are normal and 20/20, they naturally focus at about the fifteen-mile mark ahead. This is unless there is something specific to view, but in clear blue sky at an altitude with nothing else to draw the attention of my eyes, they naturally focus at roughly fifteen miles.

We are traveling at ten miles per minute, quite a normal cruising speed. Suddenly, my eyes see something smaller than a bug smear on the windscreen far ahead in the distance. I have picked up something at fifteen miles ahead, but so small as to be unable to identify what it is or even give an estimate as to its heading or size or any-thing. It is another airliner, traveling toward us at the same speed as us—ten miles per minute. Closing speed is twenty miles per minute, which is five hundred feet per second faster than a .22 caliber rifle cartridge muzzle velocity. "Faster than a speeding bullet! Superman!"

I have forty-five seconds until it is past us, so I better find out with my eyes quickly if everything this other aircraft is doing is at another altitude or at least not on my trajectory. At approximately the twenty-five second mark, I know this is a very large aircraft, in the exact opposite direction as us, and it looks to be at least a thousand feet higher than us, maybe.

If there is conflict, it will take my brain three seconds to reconcile this. It will take another three seconds to make my arms move to the throttles and flight controls, disconnect the autopilot, and start evasive action. The mass of our large aircraft will take at least five seconds to start evasive action, even if I use violent force on the controls and another five seconds to be at a new position to barely clear the opposing aircraft.

We should miss each other by approximately five hundred feet. Close but safe. If I were not looking in that direction but rather looking to the side for cross direction traffic in my scan, it would be another several seconds before all this activity would take place and we would be very close as we passed.

This is a relationship between time and speed. To which you say, "No kidding!"

Time

Einstein decided to study this relationship, but in even quicker moments of time and speed. This is where the genius of Einstein could perceive in his mind something that evades the rest of us even if we are used to the quickness of time and speed regarding life and death, as described in our scenario above.

Einstein broke the speed down to this: When you see something, by the time your eyes see it, it has already moved, and you didn't see the small amount it moved. This amount of unseen movement is based upon the speed of the object coming at you. Think of those cartoon flip charts; move them slowly, and they are stop-and-go images, but move them quickly, and they are motion pictures.

Spacetime

Einstein calls these gaps in cognizance "spacetime". In our minds this is really the distance between the past and the future. This is how Einstein thought and worked on problems like this. This thought goes on to discover that the word "now" makes no sense. It is like asking if our galaxy is above or below the galaxy of the Andromeda. That relationship is only valid if you are on a finite plain where everything is either up or down in relation to where you are standing. Even then, the folks in Australia are standing upside down, but if you ask them, they think you either are crazy or state you are the one who is standing upside down. It is perspective.

So, there isn't an up or down in the universe, and there isn't always a before and after between two events in the universe. If you read this before reading the preamble above, you would place your index finger sideways along your lips, and making a sound through your closed lips, would run your finger up and down, creating the *blb, blb mbldb*, sounds of a complete disconnect from reality. Try it. It's fun! *Blb, blb, mbldb.*

In 1905, this disconnect took the whole world of physics by surprise. It made sense out of the disconnect between Newton and Maxwell's physics. But despite this, it is not Einstein's masterpiece. The masterpiece is the second theory of relativity in Einstein's theory of general relativity.

In this theory of general relativity, the concepts of energy and mass get combined in the same way as space and time. One is dependent on, or, is a part of the other. They are two facets of the same entity, just as "space and time" and "electric and magnetic fields" are also facets of the same entity. One of the two elements of the equation can be transformed into the other. There is only one single law of conservation, not two.

Einstein did a rapid calculation and found out how much energy is obtained by transforming one gram of mass. His resulting equation is the celebrated:

$$E = MC^2$$

The amount of energy from one gram of mass is enormous. This due to the c squared part of the calculation; the speed of light times the speed of light (C^2). This brings on the era of nuclear power.

Having published his theory of general relativity in 1915, Einstein was troubled because his theory did not agree with what was known about gravity. He wondered if the Newtonian theory of universal gravity should be reconsidered as well, to make it compatible with his own relativity.

Newton imagined a force of gravity that drew objects together, whether near or distant as in falling objects on Earth or planets orbiting the sun. Newton himself was suspicious that a force of gravity between distant planets that do not touch had something missing; something that would transmit this force had to be there also. Faraday discovered the solution two hundred years later. The force of fields of electric and magnetic forces were the vehicle that transmitted the power to attract, as we now know in gravity. Therefore, the force of gravity must have its Faraday lines also. Thus, the force of attraction between the Earth and the moon, or falling objects, must be attributed to fields. This was obvious only to Einstein early in the twentieth century.

Einstein searched his gravitational field for his cognitive understanding. He calculated, made mistakes, incorrect equations, wrong ideas, and was stressed. His complete solution was finally committed to print in the book *General Theory of Relativity.*[74]Gravity fields do not operate in space.

So, what is Space?

We with small cavity brains think of space as vacuous, nothing, empty. Drink the water from a glass, and it fills with nothing. But hold on, grasshoppers. Einstein was able to reason that space was something. For fields of gravity to undulate and move, agreeing with Faraday and Maxwell's laws, there had to be particles of matter in space just like there must be particles of air in the empty glass of water. Is your mind starting to go *oouieeooo?*

Einstein discovered that space is not different from matter. It is a real entity that undulates, fluctuates, bends, and distorts. How about them apples, Mr. Newton? This is science, not science fiction or some theology asking us to have faith in this idea. Pure science. Wow!

The Earth is immersed in a giant flexible force of matter which is the universe. The sun bends the space around itself, and now the Earth runs straight in a current of a circular inclined funnel of force like a ball rolling in a funnel or on the gamblers' wheel. There is no force at the center; it is the curved nature of the funnel of force that causes the planets to circle the sun. Everything happens because space is curved. Or to be more precise, what is curved is spacetime! But Einstein needed to prove this with math, or we are going back to the finger and lips again. Try it again. It is fun.

With the help of other mathematicians, Einstein did the math. The result, space-time, curves more where there is matter. It is half a line of physics. Nobody really believed him. Not until 1980s was he taken seriously, and only then, when one after another of his predictions became verified.

Einstein predicted that time on Earth passes more quickly at a higher altitude and thus slower at lower altitudes. Remember the twin theory mentioned earlier? This was measured and proved to be correct. With today's extremely exact clocks, it is possible to measure this, even at differences of altitudes of only a few centimeters. The science is that time is not universal, or fixed, it expands and shrinks according to how close it is to a mass which distorts time. Earth slows time that is close to its mass.

---

[74] Einstein published his theory of general relativity in 1915. In it, he determined that massive objects cause a distortion in space-time, which is felt as gravity. https://www.space.com, "Einstein's Theory of General Relativity"

More of this Science

Stars burn if hydrogen gas is available, then they die out. The material remaining is no longer supported by the force of combustion and collapses under its own weight. When it happens to a large enough star, the weight is so strong that matter is squashed down in a big way, and space curves so intensely that it plunges down into an actual hole: a black hole. One of these thousands of black holes is located at the center of our galaxy, and we can see stars orbiting around it. If they pass too close, they are destroyed by the violent force ("gravity").

Space, which is not vacuous space, ripples like the surface of the sea, and the motion is the same as the ripples of electromagnetic waves that your television receives, except they are called gravitational waves. These waves were observed on Earth in 2015,[75] when two black holes fell into each other. Einstein predicted all of this and that the universe is expanding from a cosmic event 13.8 billion years ago. Oh yeah, one other thing. Often, said is that Einstein was not a good mathematician. He struggled with math all his life. His genius was his ability to see questions and solutions in his head, not unlike Faraday. Others disagree.

Some of us are starting to align ourselves to this genius, and subjectively through selective centered egotistical thinking believe we are smart, since we are also part of the same species. It is difficult for our human DNA to accept that we are not the originator of everything or the central point of the universe, and this aligns itself to this idea that we too are capable of the advanced thinking that we see in theoretical scholars. To some of us in this category, we believe we are becoming like gods. Therefore, we have answers to the questions we used to ask God. We don't need God anymore. We don't need faith because we have science, which proves everything. Faith is no longer necessary. We are becoming complacent. Atheism and agnostic points of view are becoming more and more the norm. We are becoming egocentric, and we are the center of all knowledge. We are gods, most likely. But the key word here is *believing* and that intonates a belief system. Displaced theology! Not science, just opinion.

---

[75] The first direct observation of gravitational waves was made on September 14, 2015 and was announced by the LIGO and Virgo collaborations on February 11, 2016. Previously, gravitational waves had only been inferred indirectly, via their effect on the timing of pulsars in binary star systems. https://www.space.com, "First observation of gravitational waves"

Stop, Think, Discuss

Ego is an occupational hazard of the airline pilot and likely with other occupations that require a demonstrated degree of skill. When the pilot's ego is sufficiently large, his complacency permeates his thinking as well as other side effects, like interpersonal relationships. Complacency kills. In numerous airliner fatal accidents, the cockpit recorder records the pilot whistling in a euphoric state of complacency just before the big silence. A term was coined in the industry, "Beware of the whistler."

So, I say to those who are into "science supersedes religion" or "everything is just coincidence," beware of the whistler!

Question one: In these statements, what is the divine intention of God and what is the hand of man? Is God even involved? Is this God's creation or man's? Why was Einstein so gifted within his thinking and imagination that he is seemingly beyond human capability?

Question two: What is the *ontology* of these statements? That is, what is the actual information, free from our personal bias, of these statements? And what is the *epistemology*? That is, what is the knowledge, as learned by us, including our preconceived notions, of these statements? How does the science of Einstein and others come into their minds when the normal human condition would suggest an active mind is drawn to something finite, like sports or fashion?

Question three: Einstein's thinking is progressive. What appears to come first in his mind? Were these ideas created by God or by man?

Question four: Absolutely everything that happens has a precursor that makes it happen. Then the event happens. Could you have determined, before the event, the cause of the event? Also, could the event have been prevented if the cause had been known beforehand? In this regard, is this new knowledge an invention or a discovery? Explain the origin of this discovery or invention.

Question five: How would you describe Einstein in your own words? Try to do this in ten words or less, to see if it is possible.

Question six: Are there issues in these statements with which you disagree? Has magic pixie dust been spread too thick in this description, which has detracted from the story? Give examples.

CHAPTER 40
Big Words and Big Ideas Made Easy to Follow

Scientific language is going to impact our ability to keep our minds on track with the discernment of our science. Just as doctors, lawyers, airline pilots, religious leaders, and other occupations need to express themselves in a vernacular that gives expression quickly and accurately to thoughts or actions, the same is true for scientists. They have developed language to express their thoughts to other scientific brain cavities quickly and accurately, without dancing around the subject.

Entropy

One of these words is entropy. Entropy describes a systematic progression into decay. Fire is the best example of this decay. The burning element is broken down into its elemental components, and the original source no longer exists. A melting ice cream cone is also an example, as is energy dissipating in a nuclear explosion.

The whole study of science is entropy. The disbursement of energy originally imparted upon the whole universe by the big bang at the conception of the universe is being dissipated throughout time as the universe ages. Every random activity is an action to disperse energy. Sometimes the action is random, unexplained, and unique. Other times the activity coalesces into processes that quicken the process of entropy. We *Homo sapiens* are an example of organized entropy. So is most of the science we would discuss or discover.

The moon circles the Earth and with it, the oceans rise and fall, creating predictable tides. The action of water molecules being agitated as they move takes energy. The gravity force field of Earth against the moon is affected with the equal and opposite force of Earth and moon; Newton's third law of energy is alive and well.

The result is there is less energy available the next time the moon circles the Earth because of the energy spent moving water particles. The conclusion is that the moon slowly moves away from the Earth with each orbit. The same energy loss is evident for every orbiting body in the whole universe. The amount of distance that the orbit is lengthened is obviously minuscule, but not to be ignored, as it can be measured.

So, scientists need to discuss minuscule bits of matter and time, so small that they are beyond our imagination. But they have made a whole science of small things and forces affecting our everyday lives. Very small matter is called quanta, and calculations of very small things are called quantum. We clever wordsmiths, however, not understanding the scientific brain, coined the term quantum leap, thinking we were brilliant. What we didn't know was that this would have been the smallest step ever taken by any living thing in the complete history of living things.

Quantum Mechanics

Quantum mechanics is born. Max Planck was discussed in a previous chapter. He was born in 1858, just a little earlier than Einstein, and would have been a compatriot of his. He is the fellow who discussed the unit of time we would call the nanosecond. Okay, and some of us would say that if you live in certain regions of the world, you would count to ten by using your fingers and go one, another one, and another one, and so on until you reach ten. We *Homo sapiens* have a certain disconnect from the futuristic thinking of brilliant scholars, which is proved by the thinking of Max Planck and his contemporaries. The rest of us conserve our brain's energy to only find fault and criticize. This probably further proves Darwin and Mendel's theories of mutation. We are not all equal.

> The two pillars of twentieth century physics, general relativity and quantum mechanics, could not be more different from each other. General Relativity is a compact jewel. Conceived by a single mind, based on combining previous theories, it is a simple and coherent vision of gravity, space, and time. Quantum Mechanics, or quantum theory, on the other hand, emerges from experiments during a long gestation over a quarter of a century, to which many have contributed; achieves unequaled experimental success; and leads to applications

that have transformed our everyday lives; but more than a century after its birth, it remains shrouded in obscurity and incomprehensibility.

Carlo Rovelli,[76]

Theoretical Physicist In 1900, Max Planck tried to compute the number of electromagnetic waves in an imagined hot box. This hot box was a place, in his mind, where an electric field was in equilibrium. Then, his imagination broke this field down into imaginary packets of energy (quanta). Now, this energy packet is measured mathematically and discovered to be in equilibrium when it was previously believed to be variable. Now, who the heck is going to even understand this (just trying to describe this experiment is very difficult), never mind contemplate the results of this equation? Ninety-nine-point-nine percent of the population of the universe to this day could care less. Only he could possibly explain where his mind was at, and only a psychologist could explain to the rest of us that he was not "over the edge a little."

Oh well, with the grace of understanding (maybe), we continue. Max contemplated that the energy in the field is distributed in very small packets, depending on the frequency or color of the electromagnetic waves. His equation, like that of Einstein, is very simple in the end. The letter h is a constant that he had to use to make the equation work, and it is now and forever known as the Planck constant.

Planck you, very much Mr. Planck.

Einstein finally made sense out of this in 1905, and through his work, Einstein made the discovery that light is made up of very small grains, particles of light. Aha! I see the light! He proved it by noting that there are substances that generate a weak electric current when struck by light. Today, they are photoelectric cells. A further investigation revealed that this happens more vigorously when the frequency of light is high—meaning dependent upon the color of light. For this, Einstein was awarded the Nobel Prize. We now call these grains of light packets, photons.

The more things change, the more they stay the same. I don't know who said this but, in this case, it is true. Do you remember Democritus stating that everything is made up of grains way back about 460 BCE? Hmmmm.

---

[76] *Reality Is Not What It Seems* (2017) by Carlo Rovelli, and *Seven Brief Lessons on Physics* (2016) by Carlo Rovelli. He works mainly in the field of quantum gravity and is a founder of loop quantum gravity theory. He has also worked on the history and philosophy of science.

Oh, those Danes again! My Danish father used to say there were ten lost tribes of Israel, and we never discovered who they were, but Danes know that one of them was the Danes. Now that takes faith to believe, not science. So, out of the school of science in Denmark, which has been around now since the mid-1500, came Niels Bohr.[77] He was another contemporary of Einstein. He studied the structure of atoms. Atoms make up the mass of everything, including you and me. What is not directly stated but is naturally seen is that mass has color. My telephone is black. My socks are gray. FYI, color is the frequency of light photons. But why are they not just white (all colors combined)? Why do they emit the photon frequency of black or gray? With this, let's get into the mind of Bohr. Bohr made a hypothesis that electrons only exist at certain special distances from the nucleus of an atom (i.e., only on certain orbits and on a scale that is determined by Planck's constant, h). Further, he proposed that electrons leap from one orbit with the permitted energy to another. Leaping lizards, these are the quantum leaps that we discussed earlier. Bohr-ing! The frequency that the electron moves in these orbits determines the frequency of emitted light, and since only certain orbits are allowed, only certain frequencies are emitted. Now you see the light!

Bohr defined the atomic model, which became an amazing success in experiments from then on. Munching away on Danish pastries in Bohr's Institute in Copenhagen was a German kid named Werner Karl Heisenberg,[78] who wrote the equations of quantum mechanics. Twenty-five years old, his mind was on the scale of Einstein and Faraday, where he imagined his problems and solutions. He saw a person in the dark move from one streetlight to another and then disappear. Of course, he knew that this person did not disappear. But how does this work with electrons? Did you make that connection? Probably not! He thought, what if the electron only is manifested when it collides with something else, and between collisions it has no precise position? Oh heck, I thought of that! …Not!

---

[77] Niels Henrik David Bohr was a Danish physicist who made foundational contributions to understanding atomic structure and quantum theory, for which he received the Nobel Prize in Physics in 1922. Bohr was also a philosopher and a promoter of scientific research. en.wikipedia.org

[78] Werner Karl Heisenberg was a German theoretical physicist, and one of the key pioneers of quantum mechanics. He published his work in 1925 in a breakthrough paper. He is known for the Heisenberg uncertainty principle, which he published in 1927. Heisenberg was awarded the 1932 Nobel Prize in Physics "for the creation of quantum mechanics". en.wikipedia.org

So, Heisenberg got busy calculating furiously and determined a theory. Oh yeah, another theory. I get the picture; the more things change, the more they stay the same. Change is inevitable. His theory is the first cornerstone of quantum theory. Position of particles can only be described by their positions when they interact with something else. When nothing disturbs it, an electron does not exist in any place. Heisenberg created tables of possible interactions of electrons and the tables worked. They were exactly what was seen when they were observed. These equations became the fundamental basis of quantum mechanics. To this day, they have never failed.

Stop, Think, Discuss

If you are like me, and you got to here with your eyes open and brain engaged, but just barely able to understand the simple overview, take a moment to think on a different level than the math and physics of these brilliant scholars.

When your mind works in visions like these scientists, can you see how they sometimes get their thoughts into a pattern of "everything is coincidence, not by design" or at least deist, by a creator far superior who figured it all out and it is fixed forever in rules and laws?

Can you see the process these men had, where they constantly tested through investigation supposedly fixed conclusions to determine further changes to this previously finite data?

Question one: In these statements, Planck worked in a world of the minuscule with giant thoughts. Was this God-inspired? Does he work with original thoughts uninfluenced by prior cognition? Is God even involved? Are his ideas God's influence or his alone?

Question two: What is the *ontology* of Planck's science? That is, what is the actual science, free from our personal bias, of his findings? And what is the *epistemology*? That is, what is the knowledge, as learned by Planck, including preconceived notions, of this science?

Question three: Where did Planck begin with his thinking? Was it an idea inspired perhaps by God or by man?

Question four: Absolutely everything that happens has a precursor that makes it happen. Then the event happens. Bohr redefined the atomic model. It is not possible to see the atomic model with the eye, even using futuristic instruments. Could he have determined before this insight, a cause for him to even be interested in this subject. Also, had Bohr taken up sports instead of science, would the world have missed out on this revelation?

Question five: How would you describe the events chronicled above in your own words? Also, give your opinion as to why, at this time in our history, we have arrived at the knowledge of these events?

Question six: What do you see as the personalities and insight of all these men of scientific discovery? In the descriptions of events, has Harry Potter— creative thinking—taken over the dialog on this issue?

CHAPTER 41

Another Prince of Perfection

Leaping into this ballet of perfect minds comes another young Prince of perfection. Well, maybe not perfection in the personality department? This time an Englishman named Paul Dirac[79] (1902–1984), who was a physicist thought to be the greatest physicist after Einstein.

He was possibly a savant, which means autistic as well as clever. Certainly, he had a personality that was "interesting". This acceptance of a savant presents the ultimate promise of democratic living and thinking. In some cultures, he would be an outcast, perhaps even held from doing his creative genius by the nationalism of that country. He would never be allowed to postulate any theory under autocratic rule, especially one that could possibly create stress for the leader of that society. It is only in a democratic society where everyone has sway that he has credibility and a possibility of gaining an audience.

> *... and yet in his hand's quantum mechanics is transformed from a jumble of intuitions, half-baked calculations, misty metaphysical discussions and equations that work well but inexplicably, into a perfect architecture: airy, simple, and extremely beautiful.* (Rovelli 2017)

Paul Dirac confined himself to mathematics and physics. He sorted out the confusion of others whose great thoughts create great equations. He came up with a spectrum of variability. For instance, he gave the value of the next interaction of the

---

[79] Paul Adrien Maurice Dirac OM FRS was an English theoretical physicist who is regarded as one of the most significant physicists of the 20th century. Dirac made fundamental contributions to the early development of both quantum mechanics and quantum electrodynamics. He formulated the Dirac equation describing the behavior of fermions and predicted the existence of antimatter. Dirac shared the 1933 Nobel Prize in Physics with Erwin Schrödinger "for the discovery of new productive forms of atomic theory". He also made significant contributions to the reconciliation of general relativity with quantum mechanics. en.wikipedia.org

electron in its orbit around the atom. He used probability tables. Chance exists at the atomic level, creating this probability of events.

Thus, Dirac's quantum mechanics are a way of calculating a spectrum of variables and a recipe for calculating probability that one or the other value in the spectrum will appear in the interaction. What happens between one interaction and the next does not exist. So, an electron is just a collision of particles and not an orbit of matter like most of us have been educated to believe. It is a mist or a cloud within the atom, and occasionally, it interacts with the other particles, creating a spark.

His quantum mechanics describe perfectly the structure of the periodic table of elements. Chemistry, infinitely variable in the way it combines the elements, is described in its results by quantum mechanics and this periodic table of correct data.

Quantum mechanics today offer an effective way of describing nature. The world is not made up of fields and particles, but rather a single entity, the quantum field. There are no longer particles of matter that move in space with the passage of time, but quantum fields whose elementary events happen in space.

I know right now you are thinking: *I was born but not last night!* Well, you are in a league with Einstein when you think that. He had great difficulty reasoning this. He and the Copenhagen group were in a discussion of conflict over this for a long time. Bohr and Einstein would debate this at length.

It describes everything as a relationship to something else. An aircraft is flying, and it has an airspeed, which is the relationship of the aircraft to the air surrounding it. This is not to be confused with the ground speed, which is the relationship of the aircraft to the ground over which it passes. They are two distinct and different things. But each of these relationships in the aircraft have distinct values. One is the value at which the aircraft will stall and plunge back to the Earth, the other is how quickly you will reach your destination. But somewhere in there, you do not exist! Hmmmm.

Stop, Think, Discuss

The equations of quantum physics are used to this day by numerous occupations. Medicine, biology, engineering, and chemistry are a few. The calculations work. Why? Is it coincidence? Chance? A blunder of magnificent proportion?

Perhaps it is time for a discussion about theology in science? This is different from theology is science, a discussion where mystery is the only valid reasoning, a time whereby faith is required to understand the impossible, which has not yet been defined by science. Could the quantum or quanta also be described as a God effect?

Question one: In the statement above— "the world is not made up of fields and particles but rather a single entity"—science is the focus of the statement, but what if there is another, more obscure intention? What is the divine intention of God and what is the myopic vision of man? Is God even involved? Is this God's creation or man's?

Question two: What is the *ontology* of this science? That is, what is the actual information, free from our personal bias, of these sciences? And what is the *epistemology*? That is, what is the knowledge, as learned exclusively by us, including our preconceived notions, of these statements? With the questions above, this is getting deep in thought!

Question three: What appears to come first in this? Was it an idea created by God or by man?

Question four: Absolutely everything that happens has a precursor that makes it happen. Then the event happens. Give your suggestion as to the precursor for this science to be investigated at this time.

Question five: How would you describe the idea of quanta, in your own words?

Question six: Are there issues in these statements with which you disagree? Has Harry Potter— creative thinking—taken over the dialog on this issue?

CHAPTER 42

Futurethink Paradox

There is a "pair of ducks" in the discussion of this science; or, rather, a paradox. We need to go into the future, and since *Homo sapiens* cannot know the future, but can only think of/about it, the closest we can get to the future is through faith. And so, we enter an era in science that I would call future think.

Both the theory of general relativity and quantum mechanics give us the tools that work in our world today. Almost all engineering and science use one or the other, or both, in their daily machinations. Both work and excessively well.

But there is something wrong. They appear to contradict each other. It has been stated that a university student attending classes on relativity in the morning and then quantum mechanics in the afternoon would think his professors are fools who have not recognized each other for at least a hundred years.

Both theories work very well in practice, so why don't they work together in theory? Einstein states that space and time are the manifestation of a physical field, and Bohr and company say that physical fields have quantum character. Space and time must also be quantum entities, possessing strange properties. Other Scandinavian countries would just say "those crazy Danes," but the truth is; both theories work.

What is quantum space or quantum time? Today it is a problem described as quantum gravity. It's at about this time you could say, "Oh no, gravity sucks!" And who knows, you may be right?

There is a whole potpourri of those who are working on quantum gravity to this day. It is not well conceptualized and would require a complete change in our view of the reality of our world. At one time, there were many thinking about worm holes,[80]

---

[80] A wormhole (or Einstein–Rosen bridge) is a speculative structure linking disparate points in spacetime and is based on a special solution of the Einstein field equations solved using a Jacobian matrix and determinant.

and even those who thought the discovery of the smallest particle in the atom, the Higgs boson,[81] would impart exotic material to stabilize a worm hole through which one could pass. Unfortunately, the discovery of the Higgs boson using the large hadron collider in Europe in 2013 did not help this idea.

Current thinking on quantum gravity has become loop quantum gravity. The author I have quoted more than once, Carlo Rovelli, is a proponent of this physics. He is an active player in the investigation of loop quantum gravity, using his research in quantum physics. I have read his material eight times now and revel in his ideas. The mathematics is now beyond me, even though I did study calculus at a fairly high level in my aeronautical engineering, a very long time ago. I am not a scientist, do not do research, and just read available information published by those who really want the rest of humanity to understand their perspective on everything.

Rovelli would like you to visualize the whole vastness of the universe as a giant ball of something, like a balloon full of hydrogen. After the big bang, this ball started to expand and will continue to do so for a very long time. Somehow, through entropy, the universe will start the slow process of redacting again. It will get smaller and smaller until it is back to the plasma field it was when the big bang occurred, where-upon it will commence to be a big bang again and the whole process will start all over.

If that were the case, and going beyond Rovelli and company, who is to visualize that this universe is the only universe? Could this universe be just one of many? More important, how did this all come to pass? What started it, and is it necessary to take God out of it or leave God intact in our scenario?

What if we just took out the part that we humans put into the theology, and that man invented stuff, and allowed ourselves to think way beyond the (good) metaphorical statements of religion? This is clearly what our scientists are doing

---

A wormhole can be visualized as a tunnel with two ends, each at separate points in spacetime (i.e., different locations or different points of time). Wormholes are consistent with the general theory of relativity, but whether wormholes actually exist remains to be seen. A wormhole could connect extremely long distances such as a billion light years or more, short distances such as a few meters, different universes, or different points in time. en.wikipedia.org

[81] In mainstream media the Higgs boson has often been called the "God particle", from a 1993 book on the topic. The Higgs boson is an elementary particle in the Standard Model of particle physics, produced by the quantum excitation of the Higgs field, one of the fields in particle physics theory. It is named after physicist Peter Higgs, who in 1964, proposed the mechanism which suggested the existence of such a particle. Its existence was con-firmed in 2012 by the ATLAS and CMS collaborations based on collisions in the particle accelerator in Europe. See en.wikipedia.org. "Higgs boson"

with great success, and perhaps there really is a tie-in with a reality of otherworldly intervention or input?

What our science sees is four percent of the visible universe. The visible universe is five percent of the matter that we know exists within it. We are a long way from all knowledge or all knowing.

We are not gods. We are searching for the truth, and the truth is not in us. Nor is it in those who say it is within them, for the truth is unknown. Unknown to all mankind until the truth is revealed, and not before; we have faith. Faith is our only hope into the future, and that future promises to be good. With faith and the truth revealed during history, we will have true understanding of life and our purpose within it. Then and only then will we know why.

Stop, Think, Discuss

In the big picture of the whole universe, we seem to be insignificant, and certainly not at the center of everything. And yet we are important, have future thought, investigative integrity, and seem to have a purpose in this giant play.

So why is that? Can this be explained exclusively by using science or perhaps theology, or do both require input to your explanations to make them valid?

EPILOG

The footpath of life is paved with flagstones of different types but the same origin. One stone is theology and the other is technology or science. Each of these stones is milled from the same quarry. Each is valid, each in relationship to the other, with the connecting relationship being the discourse of philosophy.

This informational footpath holds the building blocks of human life itself on this planet. From the time man evolved several million years ago to the present day, this pathway has been the foundation of knowledge and understanding. It has also precipitated the ability of *Homo sapiens* to live in harmony with each other, to some degree, without wiping each other off the face of this planet. After all, the other hominoids that shared this planet in the early era with *Homo sapiens* are long gone thanks, in large part, to the efforts of *Homo sapiens*.

We have determined, through science, that all *Homo sapiens* are related to one another. Our differences are only cultural. Culture is created by religious beliefs, lifestyle, climate, politics, food sources, and so on. This culture has set us apart and at the same time tied us together.

Our various footpaths have diverged. Some have used more theological flagstones than scientific stones. More recently, there have been those that have used more scientific stones than theology, even building entire patios out of exclusively theological or scientific stones. Those that built out of a predominance of one or the other stone have found that there is an imbalance in their culture, even failure.

The footpath of life needs balance. The Dark Ages and theocratic nations vie to reveal the failure of this balance when not using enough scientific stones, and communism is a testament to what happens when not enough theological stones are used. Tribal conflict, failed aristocratic kingdoms, and dictatorships are further testaments to this.

The only success, to date, is the pathway leading to the mutual agreement of all, and that is democracy to this point. However, this too appears to be failing, with nothing to replace it. The pathway will surely continue, but what will it be? How will we, the peons of humanity, be a part of the direction and balanced construction of this new pathway?

We are faced with the leading proponents of democracy—the United States and Great Britain–leaving their hundreds of years of proclamations of equality, liberty, justice, and fair treatment for all, in favor of recognizing cultural differences as primary in their decisions of constitutionality. Other than democracy, there is no other system of accommodating community that has survived through all of history. The footpath of life has reached a chasm. Where will it go? How will it be constructed? Will it survive the new direction? Will theology and/or science influence the pathway, and with what emphasis?

The only power of commitment to ensure that our footpath achieves a balanced direction is through us, the lowest common denominator. We alone will determine what our planet's future is for us. To do that, we have to become discerning and more understanding of each other, of science, of our structures, of theology, and of where and what we want to achieve out of all of it. We must learn from history because we cannot see the future and learn from it.

So, as you ponder our future and your role in it, learn to question, to understand, and to decide what our future will be.

This book is an effort to give you some understanding of how to reach a decision about our future. It provides you with historical information and questions to give you the impetus to discover your own questions and resolve them. Theology, philosophy, and science are symbiotic, conjoined one to the other—the pathway of life built in proper alignment and constructed with a proper foundation.

On our pathway, we need to be able to ask questions, not to be given answers that can't be questioned.

We now know that in the construct of the universe we are significantly nothing in comparison to the vastness of what we mostly don't even understand. Yet we are everything. Why?

To learn about the pathway of life, we must not only learn by reading, but must also think. Along this path, thinking is always done best through discussion. This book is a tool to start your discussion with thoughtful and historical information.

BIBLIOGRAPHY

Breur, Sander. 2017. "The Xenon Experiment." *Science.purdue.edu.* May 18.
    Accessed December 18, 2018. https://science.purdue.edu/xenon1t/?p=813.

Francis S. Collins. 2007. *The Language of God: A Scientist Presents Evidence for
    Belief.* Toronto: Simon and Schuster.

Howell, Elizabeth. 2018. "What Is The Hubble Constant?" *Space.com.* August 18.
    Accessed December 15, 2018. https://www.space.com/25179-hubble-constant.
    html.

Juliette Bently, Teacher. 2016. "Islamic sacred texts." *Religion Facts.* November 21.
    Accessed December 14, 2017. www.religionfacts.com/islam/texts/ rlft.co/1457.

Macauliffe, Max Arthur. 2008. *The Sikh Religion.* London: Forgotten
    Books. Accessed June 07, 2019. https://www.sikhiwiki.org/index.php/
    Max_Arthur_Macauliffe.

Miller, J.R. 2012. "Residential Schools in Canada." *The Canadian Encyclopedia.*
    October 10. Accessed September 9, 2019. https://www.thecanadianencyclope-
    dia.ca/en/article/residential-schools.

Quotes, James D Watson -. 2007. "goodreads.com/author/quotes/14313.
    James_D_Watson." *Goodreads.com.* January 01. Accessed December 14,
    2018. https://www.goodreads.com/author/quotes/14313.James_D_Watson.

Redd, Nola Taylor. 2019. "What Is Dark Matter?" *Space.com.* july 19. Accessed
    August 14, 2019. https://www.space.com/20930-dark-matter.html.

Rose, Devin. 2016. *The Protestant's Dilemma:*. El Cajon, California: Catholic
   Answers Inc. Accessed November 20, 2018. https://thosecatholicmen.com/.

Ross, Rupert. 2014. "Indigenous Healing." In *Indigenous Healing, Exploring
   Traditional Paths*, by Rupert Ross. Penguin Canada.

Rovelli, Carlo. 2017. *Reality is Not What it Seems.* Italy: Penguin Random
   HOuse 2017.

Srinivasan, Amrutur V. 2011. *Hindu Gods and Goddesses.* Connecticut Valley:
   Wiley Publishing.

Wells, Jonathan. 2006. "Jonathan Wells, The Politically Incorrect Guide to
   Darwinism and Intelligent Design." *Goodreads.* January 01. Accessed December
   14, 2018. https://www.goodreads.com/author/quotes/52728.Jonathan_Wells.

Wikipedia. 2018. "Baryonic dark matter." *Wikipedia.org.* December 01. Accessed
   December 22, 2018. https://en.wikipedia.org/wiki/Baryonic_dark_matter.

—. 2019. "Cepheid variable." *Wikipedia.* September 25. Accessed October 15,
   2019. https://en.wikipedia.org/wiki/Cepheid_variable.

CPSIA information can be obtained
at www.ICGtesting.com
Printed in the USA
LVHW061915131020
668697LV00018B/413